Hard SELTZER,
ICED TEA,
KOMBUCHA,
and CIDER

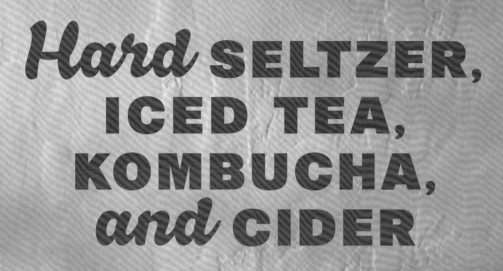

Hard SELTZER, ICED TEA, KOMBUCHA, and CIDER

How to make your own boozy fermented drinks

Emma Christensen
Photography by Erin Kunkel

TEN SPEED PRESS
California | New York

Introduction

Even now, after more than fifteen years of doing homebrewing and fermentation projects, I still feel like the transformation of relatively bland sugary liquids into tasty, effervescent, boozy beverages is pure magic. It's even more amazing that you can do this right in your own kitchen with minimal equipment and just a handful of ingredients. There is no greater satisfaction than popping the cap off a bottle of something you made yourself, hearing that happy *pffft!* of carbonation, and taking your first sip.

The fact that you are holding this book in your hands and reading these words tells me that this magic, this amazement, and this satisfaction are all things that you crave for yourself. Welcome!

With the right equipment and a basic understanding of fermentation, you can absolutely make your own homemade hard seltzers, hard iced teas, hard kombuchas, and hard ciders at home. And I'm not talking about sketchy stuff brewed in a bathtub that tastes like nail polish remover. I'm talking about real, honest-to-goodness, flavor-packed drinks that taste just as good—if not better—than what you can buy at the store.

Are you ready? Let's do this!

HOW WATER, SUGAR & YEAST BECOME ALCOHOL

All alcoholic drinks, store-bought or homemade, start the same way: combine some yeast with a sugary liquid and give it time. Yeast eats the sugar in the liquid and gives us alcohol and carbon dioxide in return. The alcohol stays in the liquid (yay!), and the carbon dioxide mostly bubbles out (at least until we trap it for carbonation purposes, but that comes later).

As long as it's a true sugar and not an artificial sweetener, any kind of sugar can be used to make fermented beverages, from honey to the corn sugar we'll use for most recipes in this book. The amount of sugar we add to the liquid will determine the alcohol by volume (ABV) of your finished drink.

As for the liquid, we'll be focusing on four main types in this book:

- Plain water to make hard seltzer
- Plain water plus tea to make hard iced tea
- Kombucha to make hard kombucha
- Apple juice to make hard cider

Some of the flavor of your finished drink comes from the sugary liquid base you start with, but a lot also comes from the particular flavorings—such as fresh fruit and fruit juices, herbs and spices, and other ingredients—you add later in the process. The flavoring stage is when the fun really starts; the possibilities are truly endless.

Steep those flavoring ingredients for a few days, then strain them out, transfer the liquid to bottles, and a few weeks later, we have a fizzy, boozy drink ready to enjoy.

THE RECIPES IN THIS BOOK

Each chapter focuses on a different type of beverage—hard seltzers, hard iced teas, hard kombuchas, and hard ciders—each with a range of recipes designed to take you from the simplest flavor options to more complex possibilities. You can choose to go in order and build your skills as you go, or you can jump around and make whatever catches your fancy.

All the recipes in this book are designed to make about 1 gallon, and they all use the same basic set of equipment as well as many of the same ingredients. I'm a huge fan of 1-gallon batches versus the more common 5-gallon batch size of many homebrewing projects, especially if you're new to homebrewing. For one thing, it's less of a commitment, both to make and to consume. If you mess something up or if the flavor doesn't turn out quite as delicious as you were hoping, it's no big deal. It's a lot less heartbreaking to dump a gallon of something down the drain than to dump larger batches. One gallon makes about ten 12-ounce bottles, which is the perfect amount for a noncommittal person such as myself. I can always make more if I fall in love with a particular recipe, but being stuck with 5 gallons of something mediocre feels daunting.

I also like that 1-gallon batches are more easily managed. This is especially great if you live in a smaller space or if you have a partner or roommate who isn't quite as enthusiastic about your new hobby as you are. Between batches, the equipment is easily tucked away in a closet or under a bed.

HOW MUCH ALCOHOL IS IN HOMEMADE DRINKS?

The exact amount of alcohol depends on the recipe itself and how much sugar is in the liquid base. The majority of the recipes in this book fall between 4.5 and 7% ABV, based on averages calculated throughout my process of working on them.

Seltzers and iced teas are on the lower end, while kombuchas and ciders are on the higher end. When you start feeling more comfortable, you can tinker with the alcohol level in your batches by adjusting the amount of sugar, but to start, I recommend following the recipes.

IS THIS SAFE?

Yes, making your own alcoholic beverages is very safe, even if this is your first time doing it, and even if you mess a few things up. Any truly harmful bugs or bacteria can't survive the fermentation process. Alcohol itself also acts as a preservative, which is at least one of the reasons humans have been happily making and consuming fermented beverages for centuries. Even beverages that are only slightly fermented are safe.

This isn't to say that your homemade drinks are immune to spoilage or infection from bacteria, mold, or wild yeast. That can still happen, and you should definitely still exercise good judgment if you notice any unpleasant flavors, aromas, or textures. But if you accidentally drink something that's off, you can rest assured that the worst you'll likely suffer is an upset stomach and not anything worse.

IS THIS LEGAL?

Happily, yes! Homebrewing and making fermented beverages at home is legal across the entire United States (check your local rules if you live outside the US). It's also fine to share or gift your homemade creations, but there are laws around selling it. If you get so excited about your new hobby that you're ready to turn it into a business, dig into your local laws and regulations before setting up a storefront.

READY TO MAKE SOME TASTY DRINKS?!

I've coached hundreds of nervous first-time homebrewers and fermenters over the years, and I've learned that the biggest key to success is pure enthusiasm. The process can seem a little daunting at first and you might worry about messing something up, but if you want it, you can make it! As long as you're excited for it, you stay curious, and you're ready for an adventure, you can't go wrong. Your new favorite hobby awaits!

Ingredients

The flavor and quality of your hard seltzer, iced tea, kombucha, or cider are 100 percent dependent on the flavor and quality of your ingredients. That said, don't feel that you have to embark on a quest for perfect, local, seasonal, artisanal ingredients to make a good-tasting beverage. Just use the best ingredients you can find and afford, and I promise those will do the job just fine.

I buy all my fermentation-specific ingredients, like yeast and corn sugar, at my local homebrew store, from an online homebrew store like MoreBeer! or Northern Brewer, or elsewhere online. I pick up grocery items like fresh fruits, frozen fruits, and fruit juices at my local farmers' market, my local grocery chain, Whole Foods, or Trader Joe's. All other ingredients can be easily found online. Check out the Resources on page 169 for a full list of all the places I shop for supplies.

BASE INGREDIENTS FOR HOMEMADE DRINKS

Think of the ingredients in this section as your "base" ingredients. They do all the heavy lifting, and you'll use them again and again for all the recipes in this book.

Water

In general, if your water is safe to drink and you're fine with the way it tastes, it will be fine for making the recipes in this book (with the exception of hard iced tea, but more on that below). If possible, filter your water before using it for fermentation, but it's not required. I don't recommend using softened water because it can interfere with yeast production and can also affect the flavor of the finished beverage.

If you want to add a little more refinement to your fermented creations, opt for spring water, artesian water, or reverse osmosis water. The lack of hard minerals and water treatment additives in these types of water will give you a stronger fermentation and cleaner flavors in your finished drink.

The only recipes where spring water, artesian water, or reverse osmosis water is truly required is hard iced tea. The various minerals in tap water (especially if it's very hard) interfere with the flavor extraction from the tea and result in a weak, muddy-tasting brew. Using one of the other types of waters mentioned will result in a more flavorful, bolder-tasting hard iced tea. (Although kombucha is a tea-based beverage, it's fine to use any type of water when making it since most of the flavor comes from the fermentation process and not from the tea itself.)

Sugar

I use corn sugar, aka dextrose, as the main "yeast food" in all my recipes. This type of sugar is not only inexpensive and widely available but can be dissolved in room-temperature water, ferments easily, and leaves very little residual flavor in the finished drink. It's a win-win-win!

Regular granulated sugar can be substituted; however, you'll need to dissolve it in hot water and let it cool before adding the yeast. Granulated sugar can also leave a faint apple- or cider-like flavor after fermentation.

Other sugars, like honey, maple syrup, agave, and brown sugar, can also be used to make fermented beverages, but they are significantly more expensive and will also leave residual flavor. I rarely use them as the main sugar in my drinks, but I like them for accent flavors, like in the Apricot-Honey Hard Cider (page 150) and the Fancy-Pants Imperial Cider (page 166).

All alcohol percentages in this book were calculated using corn sugar; substituting other sugars will alter the ABV slightly. Also, note that sugar substitutes and artificial sweeteners like stevia and Splenda are not fermentable and should not be used as the main "sugar" in your recipe. (They can, however, be used to sweeten beverages after fermentation is complete; see page 35.)

Yeast

I use champagne yeast for the majority of the recipes in this book. Champagne yeast is very easy to work with and, similar to corn sugar, doesn't leave a lot of residual flavor in the drink. I typically use Lalvin EC-1118 or Red Star Premier Cuvee.

If you're in the mood for experimenting, there is a whole world of yeast at your disposal! White wine and red wine yeasts will accentuate fruity flavors in your drinks and are your best bet if you can't get your hands on champagne yeast. Beer yeasts can also be used and often add interesting spicy notes to your drink, like in the Farmhouse Saison Hard Cider (page 158), which uses a Belgian-style yeast.

Do not use bread-making yeast. While this type of yeast will work in theory, it doesn't survive well in the presence of alcohol and will make your drinks taste extremely yeasty.

Yeast typically comes in small packets, and one packet contains enough for two or three 1-gallon batches. To store leftover yeast for your next batch, tape the opening shut, place it in an airtight container or zip-top bag, and store in the freezer for up to 3 months. Unopened packets of yeast can be stored in the freezer almost indefinitely.

Yeast Nutrient

While sugar is the main food for yeast, it will be happier and ferment better if you give it some other nutrients. I use a basic yeast nutrient blend that's a straightforward mix of diammonium phosphate and food-grade urea, with instructions to use 1 teaspoon per gallon. Other yeast nutrient blends can be used (Go-Ferm and Fermaid are two common brands), but just be sure to adjust the amount you use as per package instructions.

Do not substitute nutritional yeast, which is a different product altogether. Nutritional yeast is very good sprinkled on popcorn, but not so good in fermented beverages.

Apple & Pear Juice for Cider

Any kind of apple juice can be used to make hard cider, from apples you juice yourself to the crystal-clear, mass-produced juice served on airplanes and in kindergarten classrooms far and wide. Personally, I opt for a middle ground: I like to use unfiltered apple juice from either Whole Foods or Trader Joe's. I like the flavor of this juice on its own, and I think it makes a good base for any cider you want to make.

Pear juice can also be used as a base to make cider. It's more expensive and harder to find than apple juice, so I often blend it 50-50 with apple juice. Cider made from pear juice, aka perry, has a milder and sweeter fruit flavor than cider made from apples. If you find hard cider made from apples overly tart, try using pear juice instead!

Tea for Kombucha & Hard Iced Tea

Similar to apple juice, you can get as fancy or as unfancy as you want when it comes to the tea for kombucha and hard iced tea. Yes, higher-quality tea will result in a better-tasting finished drink, but you have to weigh this against your pocketbook. I again take the middle road with my own drinks and use the 365 Organic Black Tea and 365 Organic Green Tea from Whole Foods, which is a good-quality brand but doesn't break the bank.

For kombucha, use only caffeinated teas—caffeine is an important part of the fermentation process as well as for maintaining a healthy scoby. Avoid herbal teas or any teas with flavoring extracts, dried fruits, or dried flowers, which can cause your scoby (see below) to develop mold. Also do not use tea labeled "kombucha tea," which is a different product than what we need for making hard kombucha. For hard iced tea, any tea can be used, even decaf.

For teas used for flavoring, like the rooibos used in the Mai Tai Hard Iced Tea (page 103) or the dried hibiscus used to make the Agua de Jamaica Hard Iced Tea (page 87), I either buy loose-leaf tea from TeaSource or go for a well-known brand of bagged tea like Tazo or Republic of Tea.

Scoby for Kombucha

A scoby is a floppy, blobby, beige disk that looks a bit like a wet, rubbery pancake. It's home to all the wild yeast and bacteria needed to turn sweetened tea into kombucha. The word *scoby* actually stands for "symbiotic culture of bacteria and yeast." It usually floats on top of the sweetened tea but can also sink to the bottom or linger somewhere in the middle. The stringy brown bits that hang from the scoby are clumps of yeast and a sign that your scoby is healthy and happy.

The best place to acquire a scoby is from a friend who regularly brews kombucha (a new layer of scoby will form on top of the old one with every batch). Second best is from an online source like Kombucha Kamp, Brooklyn Brew Shop, or Northern Brewer. Be sure to get a "live" scoby that comes packaged in some of its liquid. Do not get a dehydrated scoby, as they can be tricky to revive.

INGREDIENTS FOR FLAVORING

On to the fun stuff! There are a million-and-one ways that you can flavor your hard seltzer, iced tea, kombucha, and cider, and I wholeheartedly encourage you to explore all of them. Here are the main ones we'll be using for the recipes in this book.

Liquor & Cocktail Mixers

Yes, we're making our own alcoholic beverages here, but liquors, liqueurs, and other cocktail mixers are also worthy flavoring ingredients, especially when making cocktail-inspired beverages, like the Painkiller Hard Kombucha (page 134). I've found that barrel-aged liquors, like aged rum and aged tequila, are better for making these cocktail-inspired drinks since the stronger flavor can hold its own when mixed with other flavoring ingredients.

When shopping, go for a middle-shelf option. Save the top-shelf stuff for mixing real cocktails; the bottom-shelf stuff is typically too harsh-tasting for our purposes.

Acid Blend

This ingredient might sound like something the Joker would use in a trap for Batman, but it's actually a very safe and widely used additive in the food industry. It adds pure sour flavor to everything from juice to gummy candies. Look specifically for acid blends that contain a mix of citric, malic, and tartaric acid, which has a softer, more rounded flavor than pure citric acid. If all you can find is citric acid, it's fine to use that, but start with a small amount, taste, and add more if needed.

I call for acid blend specifically in the Gin & Tonic Hard Seltzer (page 72) since it's one of the main ingredients in store-bought tonic water, but you can use it in any recipe where you'd like to punch up the flavors. A little goes a long way, so add ½ teaspoon per gallon and taste before adding more.

Fresh, Frozen & Dried Fruits

When in season, fresh fruit is always going to be best. If the skin of the fruit is edible, it can be left on, otherwise peel or cut it away. Also remove the stems and seeds, and then chop large fruits into smaller pieces. If working with fresh berries, muddle them just enough to break the skins. This will allow the juices and flavor of the berries to infuse more easily into the drink.

Off-season, the frozen fruit section of the grocery store is your best friend. These fruits are generally picked and frozen at the height of their season and are nearly as good as fresh. They can also be added directly to the fermenter, no thawing, chopping, or muddling required. The flesh of frozen fruit, even berries, will collapse as it thaws, which releases plenty of juice into your beverage.

I don't often use dried fruits in my recipes; however, they can add a fun accent flavor if you're in the mood to experiment.

I also don't often use canned fruits in my recipe. Canned fruits are usually cooked, which changes their flavor, and they're also usually packaged in sugar syrup, which alters the recipe. Use canned fruits as a last resort if other fruits aren't available and drain off any sugar syrup before using.

Fruit Flavorings & Extracts

Fruit flavorings and extracts are a great option when a particular fruit is hard to come by (like passion fruit) or if it's difficult to extract good flavor from the fruit itself (like watermelon). I generally use Brewer's Best fruit flavorings, which are available at homebrew stores and online. Use roughly 1 to 2 ounces per gallon. Other flavorings or food-grade extracts can also be used; just follow package instructions for how much to add to a 1-gallon batch.

Fruit Juices for Flavoring

Any type or blend of fruit juice can be used to add flavor to your drinks. I recommend sticking to pure fruit juices and avoiding ones that have added sugar or other ingredients. You can also use a juicer to make your own.

I buy most juices at my local grocery store. Less common juices, like guava or pear, can often be found at Whole Foods, specialty groceries, or online. Ceres is the brand of specialty juices that I use most often.

Herbs & Spices

Use only fresh herbs and whole spices for flavoring your homemade beverages— never powdered or ground versions, which can over-infuse and turn your drink bitter. For best flavor with spices, make sure they've been purchased within the past year. If you can't remember when you bought it or it no longer has a strong aroma, it's probably too old to add much flavor to your drink.

I order most of my spices online from Spice House or Mountain Rose Herbs.

Oak Cubes

Oak cubes help deepen the "barrel-aged" flavor in drinks made with aged liquors like bourbon, whiskey, and rum. They can be found at homebrewing supply stores or online.

Unsweetened Coconut Flakes

Coconut is an interesting ingredient for homebrewers and fermenters! In larger amounts, coconut flakes add a pleasant coconut flavor, just as you'd expect. But in smaller amounts, they add a hint of sweetness and a silky mouthfeel with minimal coconut flavor. This is to our advantage in drinks that could use help when balancing bold flavors, particularly in the cocktail-inspired drinks.

Look for packages of unsweetened coconut flakes (they look kind of like pencil shavings). *Do not use shredded coconut*; thin, stringlike shredded coconut tends to over-extract in the drink and will also clog up your siphon when it's time to bottle.

Equipment

Time to go shopping! The equipment listed in this chapter includes everything you'll need to start making batches of hard seltzers, iced teas, kombuchas, and ciders at home. Except for bottle caps and sanitizer, everything can be used again and again. In fact, I've been using the same basic equipment for my homemade drinks for over ten years now!

In the homebrewing world, people generally either make 1-gallon batches like we're doing or larger 5-gallon batches. All the recipes in this book are for 1-gallon batches, so be sure to pick up a smaller fermenter and a smaller autosiphon. The other equipment is the same for either batch size.

This equipment is easily found at any homebrew store, like MoreBeer! or Northern Brewer, or online. You can also check local sources for used equipment, like Craigslist or Facebook Marketplace.

MUST-HAVE EQUIPMENT

All the equipment in this section is required for making quality homemade beverages. Using the right equipment will not only minimize the risk of infection or oxidation but also make these fermentation projects easier and more fun.

Fermenter

In order to make fermented beverages, you'll need a container in which to do the actual fermenting, which we'll refer to as a fermenter. This needs to be an airtight vessel equipped with an air lock (more on that below). It should also be easy to open and close so that you can add flavoring ingredients, take sample tastings, and easily bottle your drink when it's done.

We're making 1-gallon batches, so your fermenter needs to be able to hold 1½ to 2 gallons of total volume. This allows plenty of room for a 1-gallon batch to bubble and fizz during the first few days of active fermentation, and leaves enough room to add any fruit or other flavoring ingredients later in the process.

My favorite fermenter is called the Little Big Mouth Bubbler, and it can be found at many homebrew supply stores, like Northern Brewer. It's made of thick glass, and true to its name, it has a wide mouth with a plastic screw-on lid that makes it super easy to add ingredients during fermentation and to clean after bottling.

The second-best option is a 2-gallon plastic bucket with a lid, which is inexpensive and easily found at homebrew stores and online. Be sure to buy one that is labeled food-grade and that comes with a predrilled hole in the lid for the air lock.

I do not recommend using 1-gallon glass jugs or jars as a fermenter for the recipes in this book. Technically, they will work, but the lack of headspace means that liquid is likely to bubble into the air lock during the first few days of active fermentation, and there won't be a lot of space for adding flavoring ingredients.

Drilled Rubber Stopper

Most fermenters will have a coin-size hole in the lid for a drilled rubber stopper. Along with the air lock, this creates a closed system that will keep your ferments safe and protected.

Many fermenters come with stoppers, but if yours didn't, check the fermenter's specifications on sizing and pick one up as needed. Note that plastic buckets generally don't need a stopper since the air lock can be inserted directly into the lid.

Air Lock

An air lock is a clever little device that allows the carbon dioxide produced during fermentation to escape while preventing any outside debris, bacteria, gnats, or fruit flies from getting in. There are a few different models, but they all work pretty much the same: fill with sanitizer (see below) or vodka up to the fill line (indicated on the air lock itself) and insert into the fermenter.

Sanitizer

Sanitizer isn't exactly equipment, but it is definitely an essential part of our tool kit. For the best-tasting drinks, any equipment that comes into contact with your beverage needs to be cleaned and sanitized.

Any food-grade sanitizer can be used and all generally work the same way: dilute a small amount in water, then submerge all your equipment. Follow the directions that come with your sanitizer for dilution ratios and contact times.

The sanitizer I prefer to use is called Star San, which can be reused until the liquid no longer suds up or starts to look murky. I keep about 2 gallons in a food storage container and change it out every so often. One bottle of Star San will last you years.

Mini Autosiphon, Plastic Tubing (⅜ inch) & Clamp

A siphon is the best way to transfer your homemade drinks from the fermenter into other containers, like when mixing with sugar or during bottling. Using one might seem fussy, but it will help you avoid making a mess or incorporating too much air into the beverage (which can make it taste oxygenated).

There are three parts to a siphon: the autosiphon, the plastic tubing, and the clamp. The autosiphon helps you get the siphon going, the tubing directs the liquid where you want it to go, and the clamp helps you control the flow of the liquid. Read more on how the siphoning process actually works on page 31.

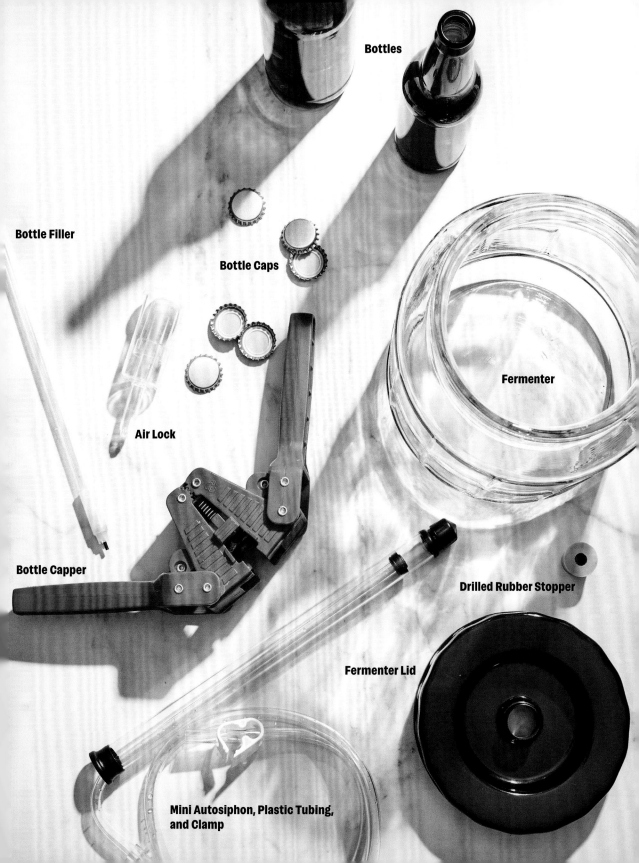

Bottles

Bottle Filler

Bottle Caps

Fermenter

Air Lock

Bottle Capper

Drilled Rubber Stopper

Fermenter Lid

Mini Autosiphon, Plastic Tubing, and Clamp

Be sure to buy a mini autosiphon; the full-size versions for 5-gallon batches are too big for our small 1-gallon batches. You'll also need 3 to 4 feet of ⅜-inch plastic tubing.

Bottle Filler

A bottle filler is a long, thin tube with a pressure valve on one end. When you're ready to bottle, you'll attach a bottle filler to the open end of your siphon's plastic tubing and insert it into your bottles one at a time. When the pressure valve pushes against the bottom of the bottle, it opens and allows liquid to flow through and fill the bottle.

Bottles

Use only bottles intended for holding carbonated beverages, like beer bottles or swing-top bottles. They will have the best seals and will also withstand the pressure buildup inside the bottle from carbonation. The best source for these bottles is homebrew supply stores.

A 1-gallon batch will fill about ten 12-ounce beer bottles, about eight 16-ounce swing-top bottles, or about six 22-ounce bottles. You can also save your store-bought beer bottles for your home fermentation projects as long as they take bottle caps and are not twist-offs. Do not buy or reuse wine bottles, as these are often sealed with corks instead of bottle caps and are generally not designed to hold carbonated beverages.

Bottles made with brown glass will provide the most protection from sunlight, and they're what I recommend using. Clear glass bottles are also fine, but be sure to keep your bottled drinks tucked away somewhere dark to avoid the sun.

Bottle Caps & Bottle Capper

Unless you're using swing-top bottles, you'll need bottle caps and a capper to seal your bottles. Bottle caps are single-use only, so I usually pick up a big bag to last me through several batches.

A bottle capper looks a bit like a butterfly or a bird in flight. By pressing down on the "wings," the device will crimp the bottle cap around the lip of the bottle, creating an airtight seal. (More on how to bottle on page 31.)

Bottle caps and cappers are universal, so buy whatever colors or designs make you happy!

NICE-TO-HAVE EQUIPMENT

The tools in this section will help you hone your craft, but you can get by without them. Add them to your tool kit as time, budget, and interest allow.

Kitchen Scale

The recipes in this book are more forgiving than other homebrewing projects, so a kitchen scale for precise measurements isn't strictly necessary. But if you want to dial in your technique or you just enjoy getting nerdy with your projects, a kitchen scale is helpful.

Hydrometer

A hydrometer measures the specific gravity (i.e., relative density) of liquids and can be used to calculate the alcohol by volume (ABV) in your recipes. I've given rough ranges for all the recipes in this book, but you can use a hydrometer to calculate the exact ABV for your specific creation, if you'd like. Instructions for how to use a hydrometer are on page 36.

Wine Thief

As the name implies, this tool is used to sneak a small taste of your beverage without exposing it to too much air or outside contaminants. Sneaking a taste is useful if you'd like to check how your recipe is coming along or add any additional flavoring ingredients before you bottle.

2-Gallon Plastic Food Storage Container with Lid

Restaurants use big storage containers, like those made by Cambro, for storing prepared ingredients in bulk. I have two in my tool kit, and I use them for storing sanitizer between batches and for mixing my recipes with sugar before bottling.

Basic Techniques

If you're new to homebrewing or making boozy drinks at home, start here! This chapter will walk you through the basic steps for every recipe in this book, from sanitizing your equipment to enjoying your first bottle.

Let's start with a bird's-eye look at the process so you know what to expect.

Day 1: Make a sugary liquid base and add yeast. Sugar is what the yeast will eat, releasing alcohol and carbon dioxide in the process. The liquid—be it water, tea, kombucha, or apple juice—is what we will enjoy drinking once fermentation is complete.

Days 2 to 6: Leave it alone. Yeast is a simple creature. Introduce it to a buffet of sugar and it will happily devour every last bit. We'll give it a few doses of nutrients to round out its meal, but other than that, yeast just wants to be left alone.

Days 7 to 14: Add flavorings. Once the yeast starts slowing down and our sugary liquid base has become a boozy liquid base, we can add some additional flavor. Fruit is always a good choice, but we can also add herbs, spices, liquor, or anything we please.

Bottle. Bottling gives us a chance to carbonate our homemade drinks. It also keeps them preserved for longer and makes them easy to share with friends and family.

Enjoy! Once all the work is done, all that's left to do is crack open a chilled bottle and start sipping.

Ready to dive into it? Read on for all the nitty-gritty details on how to transform a handful of basic ingredients into your favorite fizzy drinks.

A SPECIAL NOTE IF YOU'RE NEW TO THIS

If this is your first time doing this kind of homebrewing project, there is probably a lot that is going to feel unfamiliar and potentially anxiety-making to you. You will probably make an enormous mess in your kitchen. You will also probably wonder if you're doing it right, if you've already messed something up, and if you just wasted valuable time and money.

Trust me when I tell you that every homebrewer has gone through this. Try to think of your first batch or two as a learning experience. You might mess up a few things, but as long as you're patient with yourself and are having a good time, you'll be just fine.

From my experience, the two places where newbies get the most nervous are around sanitizing equipment and bottling the finished drink. Read those sections carefully and maybe watch a video or two online. I also recommend doing a test run bottling plain water to get a feel for how the siphon and bottle capper work.

Even better, buddy up with a more experienced homebrewer who can show you the ropes and walk you through unfamiliar techniques. There's also a troubleshooting section at the end of this chapter to help you with some of the more common issues that might come up. You've got this!

DAY 0: BEFORE YOU BEGIN

I once had a teacher who would constantly say, "Give yourself every opportunity to succeed: be prepared." It's a motto I've taken to heart, both in my life and in my brewing and fermentation projects. A successful batch starts before the yeast even comes into play: with clean equipment, good ingredients, and a plan of action.

How to Schedule Your Fermentation Projects with Your Life

All the projects in this book are designed to be made and ready to drink in about a month. I usually start a recipe on a Saturday or Sunday, add the flavoring ingredients the following weekend, bottle one to two weeks after that, and enjoy my chilled, fizzy beverage a week later. If this is your first time brewing, try to pick a time when you'll be home for the full four weeks so you can keep an eye on each stage.

But even with the best intentions, sometimes our lives or our ferments don't go according to schedule—an unexpected work situation might distract you for a few weeks, or fermentation might take a few extra days to complete before you can bottle. It's OK! Fermentation projects are generally very forgiving, and it's not the end of the world if things don't happen exactly on the timeline they're meant to. (For some tips on how to modify your timeline if something comes up, check out the troubleshooting section on page 38.)

DAYS 1 TO 6: START YOUR FERMENTATION ENGINES!

Today's the day! In the beer world, we call this brew day, but given that not all the various projects in this book involve actual brewing, let's just call it Day 1. This is the day when your fermentation project begins: you'll mix together your sugary base, add some yeast, and wait in eager anticipation for the first signs of yeasty activity.

How to Sanitize Your Equipment

The day you're planning to make your recipe, gather all the equipment you'll need and give it a good scrub with soapy water. You want everything squeaky clean, from the fermenter that will hold your recipe as it ferments to the whisk you'll use to stir it up.

Add another layer of protection by sanitizing all your equipment; this helps ensure that any stray germs have been neutralized. Most sanitizers, like the Star San that I use, require you to dilute a small amount of sanitizer in water. It's easiest to do this right in your fermenter, which you need to sanitize anyway, though you can also use a large bowl, a bucket, or other big container.

Plop the equipment you'll need for the day into the sanitizer and make sure it's submerged. This includes the fermenter lid, air locks, spoons, everything. If some things aren't totally submerged, let them soak for a bit and then flip them around so that every part gets in contact with the sanitizer. Be sure to splash the sides and the mouth of your fermenter if the sanitizer doesn't reach that far.

Check the sanitizer instructions for how long equipment should be submerged to be sanitized—it's usually around a minute or two. You can also leave the equipment in the sanitizer until you're ready to use it, or you can lay everything out on clean kitchen towels to dry until you're ready to use it. (Sanitized equipment can generally be used wet or dry; just don't rinse anything after sanitizing unless your sanitizer instructions tell you to do so.) If your sanitizer is the kind that can be reused, be sure to save some or all of it to fill your air lock later or in case you need to re-sanitize anything.

How to Measure Your Ingredients

Once your equipment is ready, next up is measuring your ingredients. For most of the recipes in this book, the ingredients at this stage will be water, kombucha or apple juice, sugar, and yeast.

Measure water and other liquids in a liquid measuring cup. Sugar is best measured using a kitchen scale, but if you don't have one, scoop the sugar with a dry measuring cup and sweep the top with the flat side of a butter knife for the most accurate measurement. Yeast can be measured with measuring spoons.

How to Mix Your Sugary Base

If you were using your fermenter to sanitize your equipment, move all the smaller equipment to clean dish towels and pour out the sanitizer.

Pour the sugar into the fermenter and top it with the liquid for the recipe you're making today. Sprinkle the yeast over the surface of the liquid and let it sit for a minute or two, until the individual grains have started to dissolve and sink below the surface. Then whisk with a sanitized whisk until all the sugar is dissolved and the liquid is slightly foamy on the surface, 30 to 60 seconds.

That's it! Now seal your fermenter by securing the lid and inserting the rubber stopper (if your fermenter uses one). Fill your air lock with some of your reserved sanitizer (see page 25), vodka, or (in a pinch) filtered water up to the indicated fill line, stick it into your fermenter, and you're done.

How (and Where) to Store Your Recipe During Fermentation

The best place to store your fermenter while the yeast gets down to business is somewhere dark, slightly warm (70° to 80°F), and out of the way—though not *too* far out of the way since you'll want to keep an eye on it over the next few weeks. A hallway closet, a kitchen cupboard, or a room that doesn't get frequent traffic (like a laundry room) is great. If you need to store it somewhere in the open, like a countertop or an open shelf, wrap it in a tightly woven dish towel to minimize exposure to sunlight.

It's also good to place a towel under the fermenter. It doesn't happen often, but sometimes if the yeast is especially happy, the liquid can bubble up to the top of your fermenter and push into the air lock.

How to Tell if Fermentation Is Happening

Fermentation should kick off within 12 to 24 hours after mixing everything together. Warmer temperatures speed things up and cooler temperatures slow things down.

The surest sign that fermentation is underway is if you see bubbles popping up through the air lock. This means that the yeast is happily chowing down on the sugar in the liquid and releasing carbon dioxide in the process. This bubbling will be quite vigorous for 3 to 4 days and then start to taper off. By the end of 7 days, you should be seeing only a stray bubble or two in the air lock every so often, if any.

If you don't see any bubbles in the air lock within 12 to 24 hours, something is off. The most likely explanation is that the lid is a little loose or there's a tiny crack or tear in a seal somewhere. Try opening your fermenter and closing it again, checking to make sure that everything is sealed as tightly and securely as possible.

If you still don't see bubbles in the air lock within an hour or so, don't panic yet. Check out the troubleshooting section on page 38 for more ideas.

How to Add Yeast Nutrients

Yeast nutrients are helpful partners for making hard seltzers, iced teas, and kombuchas. (You can also add them to cider, but they're less essential.) They are precisely what they sound like: nutrients for the yeast, which has been living on an all-sugar diet and needs some extra nutrition to stay healthy. Without extra nutrients, yeast can become stressed, which can result in slow fermentation, stuck fermentation (when fermentation stops before all the sugar is gone), or off-flavors.

The yeast nutrient I use is a straightforward mix of diammonium phosphate and food-grade urea, and the instructions are to use 1 teaspoon per gallon. I like to space this out over 3 days to give the yeast a few separate snacks. Check the instructions that come with your yeast nutrient; if they're different from mine, just divide the recommended amount for 1 gallon into three doses and add them over 3 days.

Dissolve your dose of yeast nutrient in a few tablespoons of warm water (tap water is fine), open your fermenter, and pour it in. Seal the fermenter again (no need to replace the sanitizer in the air lock unless you need to top it off) and give it a few gentle swirls to make sure the yeast nutrient gets mixed in. Repeat for the next two doses over the next 2 days, and you're done.

DAYS 7 TO 14: ADD SOME FLAVOR!

The first week of fermentation gives you a boozy, alcoholic base. Now we get to have fun adding flavorings.

When to Add Flavoring Ingredients

In general, you can add your flavorings any time after fermentation slows down. If I've started a batch on a Saturday or Sunday, then I'll usually add the flavorings the next weekend (after 7 days).

Flavors from these added ingredients generally get stronger the longer they sit with your boozy base. At a minimum, I recommend 3 days. At a maximum, you can leave them for up to 3 weeks (longer than that and you can sometimes start to develop mold or spoiled flavors). I find that 1 week of infusion is typically a good middle ground for extracting the best flavor and staying on schedule.

How to Prep Flavoring Ingredients

The alcohol already in your beverage will help protect it against outside contaminants introduced from the flavoring ingredients, but it's still good to exercise common sense when handling and preparing them. If you're working with whole fruits that are easily washed (like citrus fruits or apples), wash them with soapy water and dunk them in sanitizer before adding them to the fermenter. Don't worry about washing or sanitizing ingredients like spices or herbs.

Trim off any stems, leaves, bruises, or nonedible bits from the fruit, and either chop into bite-size pieces or mash slightly to start releasing their juices. If the peels of the fruit are edible, they can be left on.

Frozen and dried fruits can generally be added as is unless otherwise instructed in the recipe. As the fruit thaws in your beverage, the cell walls will collapse and release their juices.

To remove strips of zest from citrus fruits, use a vegetable peeler and scrape off long strips from stem to end. Try to remove only the colorful, fragrant outer zest, leaving as much of the bitter white pith behind as possible. If your citrus fruit isn't zesting easily, put it in the freezer for 15 minutes; this should freeze the outside just enough to make it easier to strip away the zest.

Juice the citrus by cutting it in half and using a hand juicer or countertop juicer, or just squeeze it with your fingers. Don't worry if the juice is pulpy—it's more flavor for your recipe!

Prep fresh herbs as directed in the recipe. Herbs do not need to be bruised or muddled before being added.

All other ingredients, like spices or coconut flakes, should ideally be measured with a kitchen scale for best results, but your recipe will be just fine if you measure by volume. They can be added directly to the fermenter without additional prep.

To add any of these flavoring ingredients, open the fermenter and add the ingredients directly to the liquid. Try to splash as little as possible. When you're done, seal the fermenter back up, give it a gentle swirl to mix in your flavoring ingredients, and return it to where you've been keeping it.

How to Taste & Adjust the Flavors

While you don't *have* to taste your recipe before bottling, it's a good practice to get into. This gives you a chance to evaluate the flavors and make any final adjustments before bottling. I usually taste my beverages a few days before I plan to bottle.

The best way to sample your drink is to use a wine thief to withdraw a small taste: Insert the wine thief a few inches into your drink and let it fill. Then put your thumb over the top opening to trap the liquid in the thief, remove it from the fermenter with your thumb still in place, and release the liquid into a glass. Second best is to scoop out a little with a sanitized measuring cup or ladle.

Here are some ways you might adjust the flavors in your recipe:

- If the fruit flavor isn't as strong as you'd like, add more fruit! When I do this, I usually add half the amount I originally added, taste a few days later to see how I like it, and then add even more if I want.

- Ditto if the herb or spice flavor isn't as strong as you'd like. Just add half the amount originally added and taste a few days later.

- If the flavor seems "fine" but is a little boring or uninteresting, try adding strips of zest and the juice of 1 lemon or lime. This isn't typically enough to make your drink taste like lemon or lime, but a little citrus has a magical way of perking up other flavors.

- If you'd like a boozier flavor, add half the amount of liquor originally added. Or if the original recipe didn't call for liquor already, start with adding ½ cup of a liquor that complements your recipe and then add more from there.

- If the flavor is too sharp or dry for your liking, try adding ½ cup of coconut flakes (see page 15). This will add a touch of sweetness to your drink without adding much actual coconut flavor.
- For hard tea recipes, add more tea bags if you'd like a stronger tea flavor.

With any flavor adjustments you make, wait 2 to 3 days, taste again, and then either make further adjustments or wait for fermentation to stop before bottling.

BOTTLING DAY!

Bottling your homemade drink accomplishes a few things: it allows you to carbonate the drink (which generally makes it more enjoyable), it preserves the drink for a longer period of time, and it allows you to share your creation with friends and family.

When to Bottle

Your homemade beverage is ready to bottle as soon as it stops actively fermenting. I typically wait until I see zero signs of bubbling in the air lock, though a stray bubble or two is usually fine (especially with kombucha, which sometimes continues ever-so-slowly fermenting for longer than my patience can withstand).

Avoid bottling too early, while the drink is still actively fermenting (i.e., bubbles still popping frequently through the air lock), because this risks the beverage overcarbonating in the bottle. Overcarbonation can cause the drink to gush when you open it or—worst-case scenario—your bottles to crack. Either way, you'll have a mess to clean up and less homemade drink to enjoy.

In general, it takes about 2 weeks total for the yeast to consume all available sugar and for fermentation to complete. If you add fruit or another sugar source, fermentation will become active again and then gradually slow within 3 to 4 days. I usually add fruit and other flavorings a week before I plan to bottle so that there's plenty of time for fermentation to finish before bottling. If you're not sure whether fermentation has finished, it's best to wait a few more days rather than bottle too early.

Also, note that cool temperatures will slow fermentation. If your home is on the cooler side, fermentation may take a little longer than stated in the recipe.

How to Carbonate, Siphon & Bottle Your Drink

Before bottling, sanitize all your bottles inside and out, a large stockpot, your autosiphon and tubing, a bottle filler, and bottle caps. Attach one end of the plastic tubing to the top of the autosiphon and position the hose clamp on the other end. If the tubing is a little stiff or too tight to attach to the autosiphon, try warming it in hot water for a few minutes to make it more pliable.

Place your fermenter on a counter and place the sanitized stockpot on a step stool or chair so that it's a foot or two below the fermenter. Dissolve the corn sugar for bottling in ½ cup of water and pour this into the sanitized stockpot.

This last bit of sugar is how your drink will become carbonated. The yeast will eat this sugar while in the bottle and release carbon dioxide and a small amount of alcohol. The carbon dioxide stays trapped in the bottle, which is how the drink becomes carbonated. (The amount of alcohol produced is small enough that it doesn't significantly change the ABV% of your drink.)

Next, remove the lid from your fermenter and slide the autosiphon along one side of the fermenter until the tip rests on the bottom. Try to disturb the sludge on the bottom as little as possible.

Position the open end of the tubing in the stockpot with the corn sugar and gently pump the autosiphon a few times to start liquid running through it. Once going, the siphon should work on its own without need for additional pumping.

While starting your siphon, be careful since the open end of the tubing can jump around a bit and spray liquid all over your kitchen while you pump. I usually hold the open end of the tube with one hand while I work the autosiphon with the other. This takes a bit of coordination, so it's nice to have a buddy nearby who can help.

Siphon all the liquid from the fermenter into the stockpot, leaving behind any solid flavoring ingredients. Toward the end, you can tilt the fermenter to siphon as much liquid as possible, but take care to do this gently so you don't stir up that bottom sludgy layer too much. Once you've siphoned all the liquid, transfer the autosiphon to the stockpot and lift the stockpot to the counter.

Arrange all your bottles on the step stool or chair; I usually put them on a tray to keep them stable and catch any accidental drips. Attach the bottle filler to the open end of the siphon tubing. (If it's a tight fit, warm the open end of the siphon tubing in hot water for a few minutes.) Insert the bottle filler into the first bottle and press the tip against the bottom. Pump the autosiphon again to start the flow of liquid.

Fill until the liquid reaches the lip of the bottle, then lift the bottle filler out. As soon as you lift the filler, the pressure valve in the tip will stop the flow of liquid. Once the filler is removed from the bottle, it will leave the exact right amount of empty headspace in the bottle so the liquid can carbonate.

Repeat with the remaining bottles until all the liquid is gone. If your final bottle is at least half-full, go ahead and cap it. If it's less than half-full, I consider this a treat and pour it into a glass to enjoy myself.

How to Cap Your Bottles

Lay a clean dish towel on your counter and arrange all the bottles on top; the towel helps keep the bottles from sliding while you cap them. If you are using swing-top bottles, you only need to clamp the cages down around the neck of the bottle.

For bottles requiring a cap, set one sanitized bottle cap over the mouth of the first bottle. Position the bottle capper over the bottle so the circular part in the middle of the capper rests against the bottle cap. Press down gently but firmly on the handles (the "wings") until they're parallel with your counter. This will squeeze the capper around the bottle cap, pressing the flared sides of the cap around the lip of the bottle and creating an airtight seal.

You don't need to use a lot of force when you do this! As long as the cap is snug and there's no give, you can pat yourself on the back for a job well done.

How & Where to Store Your Bottles

Place all your bottles somewhere dark, at room temperature, and out of the way. I usually put them in a cardboard box and stash the box in a closet or in the basement. The box makes them easier to move around and also keeps any mess contained on the off chance that one of the bottles overcarbonates and cracks.

Wait at least 1 week before opening and drinking any bottles to give the liquid time to carbonate. If you open a bottle and it's undercarbonated or you notice a yeasty flavor, wait a few more days before opening any more bottles. If you've made hard kombucha, I recommend moving your bottles to the fridge after a week or two. Kombucha can sometimes continue fermenting slowly, and refrigerating helps prevent overcarbonation.

These homemade drinks are meant to be consumed fresh, but they can be stored for up to 3 months without any significant change in quality or flavor. They may last even longer depending on how well you sanitized everything, whether they were exposed to light during or after fermentation, and how well the bottles were stored. If you come across an older bottle, give it a taste! If it tastes good to you, go ahead and drink it. If not, then pour it down the drain and start another batch.

How to Clean Your Equipment

Clean all equipment thoroughly with soapy water right after you're done with it, including your fermenter, empty bottles, siphon, and all the rest. Use abrasive sponges and bottle brushes to get in every nook and cranny, and run soapy water several times through plastic hoses or tubes too narrow for brushes. Even a little gunk left behind from a previous batch could spoil your next one. Rinse thoroughly with hot water and let everything air-dry before storing it.

TIME TO TASTE!

At long last, the time has come to enjoy the fruits of your labor. Whether you're pouring yourself a solo glass after work or sharing your homemade drinks with friends and family, be sure to take a minute to appreciate this awesome drink you've made!

How to Enjoy Your Homemade Drink

All the recipes in this book are best enjoyed chilled, so put your bottles in the fridge for a few hours before opening. When ready, grab yourself a clean glass and then pop the top off your bottle over a sink or outdoors. (This is in case it's overcarbonated and starts to gush when you open it.)

No fancy pouring techniques are needed for these drinks; just pour it right into the glass! Add some ice if you'd like for longer-lasting chill, or drink as is. If you notice sediment at the bottom of your bottle, stop short of pouring this into your glass. It's fine to drink if you do, but it can make your drink taste yeasty.

Before drinking, hold your glass up to the light and note its color. Give it a good sniff to take in the aromas. Hold the first sip on your tongue for a moment so you can fully taste it. If you feel so inclined, write down a few notes so you remember what you liked, didn't like, or want to try differently the next time you make this recipe.

After that, just sit back and enjoy.

How to Sweeten Your Homemade Drink

Homemade fermented beverages tend to end up a lot less sweet than their store-bought counterparts. This is because yeast will keep eating the available sugar until it's gone, so even if you add more sugar after fermentation has stopped with the intention of sweetening the drink, the new sugar will just kick-start fermentation again. Commercially made drinks don't have this same problem because manufacturers have a lot more industrial tricks and techniques at their disposal. There are ways that we homebrewers can neutralize the yeast so that it stays dormant even after adding additional sugar, but these are advanced techniques beyond the scope of this book.

Instead, I prefer to sweeten homemade drinks by adding a spoonful or two of simple syrup to the glass before pouring the drink over top. Simple syrup is easy to make: just combine equal parts sugar and water, warm in the microwave or on the stovetop until the sugar dissolves, then let cool and refrigerate for several weeks. You can use any sugar you want, like regular table sugar, honey, maple syrup, or brown sugar.

You can also add nonfermentable or artificial sugar substitutes before bottling, like stevia or Splenda. I don't personally enjoy the flavors of these artificial sweeteners, but if you want to go this route, add a small amount of your nonfermentable sugar along with the bottling sugar and siphon the beverage over top. Taste and continue adding more until it's sweet enough for your liking.

BONUS TECHNIQUES

If you get into making homemade drinks, here are a few more techniques that can help you up your game.

How to Use a Mesh Bag to Hold Flavoring Ingredients

Many homebrewers prefer to put their flavoring ingredients in a mesh bag so that it's easier to separate them from the liquid when it comes time to bottle. This can be especially handy for small ingredients or very fine ingredients that might clog up the siphon.

You can purchase large mesh bags online or at homebrew supply stores. I usually sanitize the bag first, then add the flavoring ingredients. To close, I either knot the top or use a sanitized rubber band. You can just drop the entire bag right into your fermenter.

To clean after using, turn the bag inside out, rinse out all the solid bits under running water, and toss it in the laundry the next time you're doing a load.

How to Calculate the Alcohol by Volume (ABV) of Your Recipe

You can calculate the alcohol content of your drink fairly easily with a tool called a hydrometer, which is used to measure the gravity (i.e., density) of liquids. In our case, dissolved sugars make our drinks more dense and lack of sugar makes them less dense. If we take one measurement at the start of fermentation (before any sugars have been consumed) and another just before bottling (after the yeast has consumed the majority of the sugars), we can use the difference between these two numbers to calculate how much alcohol is in the drink.

To use a hydrometer, sanitize the hydrometer, its tube, and a small measuring cup. Put the hydrometer inside the tube, scoop out about ½ cup of your drink, and pour it into the tube until the hydrometer begins to float. Take note of where the surface of the liquid hits the hydrometer and write down the gravity somewhere handy.

The reading you take at the start of fermentation is called the *original gravity*. The reading you take at the end is called the *final gravity*. Here's the formula for calculating the ABV%:

(Original gravity - final gravity) × 131.25 = ABV%

So if your original gravity is 1.038 and your final gravity is 1.000, then your ABV is 4.98%, or roughly 5% alcohol.

How to Increase (or Decrease) the Alcohol Content in Your Recipe

One way that you can play around with the recipes in this book is to adjust the amount of sugar to create a less or more boozy drink. Less sugar will give you a lower ABV, and more sugar will give you a greater ABV.

Use a hydrometer for exact calculations since the relationship between sugar and alcohol content isn't perfectly linear, and it's also dependent on things like the type of sugar used and flavoring ingredients you add. But for general estimation purposes, you can figure that every 2 ounces (6 tablespoons) of corn sugar dissolved in a gallon of liquid will increase the gravity by roughly 0.004 and ultimately increase the final alcohol content by about 0.5%.

How to Use a Mini-Keg

One-gallon mini-kegs do exist and are a worthwhile investment if this becomes an ongoing hobby for you. They're much easier than bottling since you can siphon the entire batch directly into the mini-keg. You also don't need to add any extra sugar since the keg will pressurize and carbonate the drink for you. The only downsides are that you need to plan on drinking your entire batch within a few weeks, and it's less easily shared.

To use a mini-keg, sanitize the inside of the keg, including the spigot, the tube that holds the CO_2 cartridge, and the bottom of the cap. Siphon your entire batch into the keg up to the fill line. Follow the instructions that come with your mini-keg to pressurize the keg to your preferred level of carbonation (typically 9 to 12 psi for the recipes in this book).

How to Scale Up Your Batch

If you love a recipe so much that you'd like to make 3 gallons or even 5 gallons of it, you can absolutely do that. All the recipes in this book can be scaled up to whatever batch size you desire without needing to make any major adjustments. You'll just need a larger fermenter, a longer autosiphon, and more bottles, all of which are readily available at homebrew stores.

TROUBLESHOOTING

Although making homemade hard seltzers, iced teas, kombuchas, and ciders is a fairly straightforward process, unexpected things do crop up from time to time. Here are some of the most common questions and what to do about them. (Spoiler alert: most of my answers boil down to "It's probably OK! Don't panic!")

What if I have to modify the schedule in the recipe?

Life happens! If you need to modify the timing of any of the recipe steps, there are several moments when you have some flexibility:

- If you won't be around to add the yeast nutrient in three doses, either add it in one big dose at the beginning along with the yeast itself or whenever you're able during the first few days.

- The flavoring ingredients can be added as soon as you notice fermentation starting to slow down (around Day 4 or Day 5), or you can wait to add them for up to 1 month. (Note that hard kombucha can sometimes become tart if left this long. There will be no difference in flavor for the other recipes in this book.)

- Unless otherwise noted in the recipe, flavoring ingredients can be left in the fermenter for up to 2 weeks if you're playing it safe, or 3 weeks if you really need to push it. Much beyond that and you might start to notice over-extracted or spoiled flavors in your batch.

- You can also remove the flavorings with sanitized tongs (or siphon the batch to a fresh fermenter) and wait to bottle until you have time. Your recipe can hang out in the fermenter like this (with the flavoring ingredients removed) for as long as you

need, though the yeast will start to go dormant after 2 to 3 months.

- All of these recipes are intended to be consumed fairly fresh. Once bottled, I recommend drinking them within 3 months. They are often good for longer, but you might start to notice oxidized flavors or other off-flavors.

I forgot to sanitize my equipment!

One of the only things that can truly ruin one of your batches is if it gets infected with stray bacteria or wild yeast. Thoroughly cleaning and sanitizing every piece of equipment that comes into contact with your drink at every stage of the process is the best way to prevent this from happening.

That said, we all slip up from time to time. Yes, the risk of infection increases, but if you've kept things generally clean and sanitized, the risk is low. There's no need to toss your batch; just keep an eye on it and make sure to sanitize everything going forward.

I forgot to add the yeast nutrient!

If it's still within the first 4 days after mixing up your batch, go ahead and add today's dose as soon as you remember, as well as any you missed. If you've passed the 4-day mark, add all the yeast nutrient at once. If a full week has gone by, then skip the yeast nutrient for this batch.

The longer you wait to add the nutrient, the less useful it is for the health of the yeast. Adding it late or forgetting it entirely doesn't mean your batch is ruined by any means, but you might notice some harsher flavors than usual.

I don't see any bubbles in the air lock!

Bubbles merrily popping through the liquid in the air lock is the best way to tell that fermentation has started, but don't panic if you don't see them. First, make sure there's sufficient liquid in your air lock. All air locks have a "fill line" somewhere near the top. Top it off (or pour a little out) if needed.

Second, check the time. Has it been 24 hours? If not, wait longer. Yeast takes 12 to 24 hours to really get going, especially if it's chilly in your house. If it's chilly, move the fermenter somewhere warmer, closer to 70°F. If it's already been 24 hours, move on to the next few suggestions.

Next, try removing the lid from your fermenter and checking for cracks or tears. If you find any, replace the part if you can, or carry on if you can't. Cracks or tears can introduce the possibility of infection, but it's not a done deal. In the first few days of active fermentation, there's enough carbon dioxide being pushed out from the fermenter that your recipe will stay relatively protected. Replace the lid and the air lock and check again for bubbles in a half hour.

If you're pretty sure your lid is OK and you still don't see bubbles, take a look inside. Do you see foam or bubbles on the surface of the liquid? Do you see movement within it, like yeast particles floating up and down? Do you hear a soft fizzing noise or smell fruity, fermenty aromas? All these are signs that the yeast is doing its job and fermentation has started. There's probably a microscopic crack or a tear in the seals of your container, which is where the CO_2 is escaping, but at this point you can carry on as usual.

If you still see zero signs of fermentation, add all the yeast nutrient called for in your recipe, whisk vigorously with a sanitized whisk for 30 to 60 seconds, and make sure the fermenter is somewhere at warm room temperature. Still no luck? The most likely culprit is a bad batch of yeast. In this case, if you have another packet of yeast or can get more within a day or two, add that. If not, then go ahead and toss this batch and start again once you have new yeast.

I stopped seeing bubbles in the air lock after a few days. Is that supposed to happen?

Yes! The yeast will be very active in the first several days and then activity will slow toward the end of the week. If it's warm in your house, this might happen more quickly. Either way, everything is just fine.

My batch is still bubbling after 2 weeks!

A stray bubble or two in the air lock as you near the end of the second week of fermentation is fine. If you added flavorings toward the end of the second week, if the temperature is a little chilly in your home, or if the yeast is just extra happy, fermentation may continue into a third week, which is fine.

However, if you're still seeing a lot of rapid bubbling with no sign of slowing, you might have a problem. Open the fermenter and take a peek. If you see green or black mold, or if your recipe smells swampy or like a gym locker, it likely picked up an infection of some kind. If you're not sure, let it continue for a few more days—infections quickly become *very* obvious! If you think your batch has been infected, toss it, clean all your equipment thoroughly, and try again.

If you're making hard kombucha, you will often see a stray bubble or two in the air lock even past the second week. Since kombucha contains its own naturally occurring bacteria and yeast, it can sometimes continue fermenting very slowly for quite some time. This is fine and normal. As long as you're seeing only a stray bubble or two lingering in the air lock, it's fine to proceed with bottling.

What's the gunk on the bottom of my fermenter or floating on the top of the liquid?

As active fermentation slows, the inactive yeast will start to fall to the bottom of your fermenter, creating a sludgy layer of sediment. Likewise, some of the ingredients you add might sink to the bottom. This is all fine and normal.

You might also see gunk collecting on the surface of your batch. Sometimes this happens if fruits or other ingredients float and then get coated with a layer of yeast sludge. Hard kombucha also often collects a scuzzy brown layer on the liquid's surface.

All these things are normal. As long as you don't see fuzzy green or black patches of mold and as long as your batch smells OK, you're fine. Good smells are yeasty or fruity aromas. Bad smells are swampy, locker room, or rancid aromas.

One of my tea bags burst!

This is fine! Carry on as usual and be extra careful when bottling to avoid transferring as many of the tea particles as possible.

Some of the sediment got transferred when I bottled!

This is also fine! If a little (or even a lot) of sediment from your fermenter gets transferred when you bottle, it won't ruin your batch. In larger amounts, it can add a bitter, yeasty flavor or a gritty texture, but even still, it won't ruin your bottled drinks.

It will also likely settle to the bottom of your bottles, so when you pour, just avoid pouring the last ½ inch or so.

My finished drink was hazy!

Not to fear! Haziness is cosmetic and is caused by small particles of fruit or yeast still suspended within the liquid. This haziness does not have a significant effect on flavor. The particles will eventually settle to the bottom of your bottle, so if you'd prefer a clear drink, just wait a few weeks before enjoying.

My bottles didn't carbonate!

Did you wait at least a week before opening your first bottle? If not, then give things a few more days and try another bottle. Also, if you're storing your bottles somewhere cooler than room temperature, they will take longer to carbonate. Either wait longer or move the bottles somewhere warmer.

Did you remember to add the final dose of corn sugar for bottling? Yeast needs this last bit of sugar in order to produce the carbon dioxide that will carbonate your drink. If you forgot it, then you can either drink it flat or you can uncap your bottles, divide the sugar evenly, recap all the bottles with new caps, and wait another week before sampling.

Did you wait longer than 3 months to bottle your batch? If so, the yeast may have gone dormant and wasn't able to revive enough to carbonate your drink. You can either drink your bottles uncarbonated, or you can try dissolving a small amount of yeast in water, dividing this between the bottles, recapping with new caps, and waiting a week before sampling.

Was this your first time bottling homemade drinks? If so, then you might just need practice using your capper. If the cap isn't uniformly crimped around the lip of the bottle, air can escape and leave your drink uncarbonated. It's also possible that this bottle was a fluke; try another bottle tomorrow to see if the same thing happens!

Did you carbonate in bottles not intended for carbonation? Use only beer bottles (12 ounces or 22 ounces) or swing-top bottles purchased from a homebrewing supply store. Vessels like canning jars, water carafes with swing-top closures, and bottles without airtight seals will not work.

My bottles gushed when I opened them!

It sounds like your bottles overcarbonated slightly. This can happen if the batch wasn't quite done fermenting when you bottled it, or if you mismeasured the final dose of corn sugar.

Make sure your bottles are thoroughly chilled (which reduces carbonation) and open them over a sink as slowly as possible, releasing the pressure gradually.

A few of my bottles cracked or burst!

This is also due to overcarbonation, just to a greater degree. Basically, the yeast created too much carbon dioxide inside the bottle, pressure built up, and after a while, the bottle couldn't withstand the pressure anymore.

Refrigerate all unopened (and uncracked!) bottles immediately and be careful when opening them.

My drink isn't sweet enough or tastes too tart!

This is very common with homemade drinks since the yeast eats all available sugar, leaving nothing behind for our sugar-loving palates. You can sweeten your batches with nonfermentable sugars, like stevia or Splenda, or you can add a few spoonfuls of simple syrup to your glass when serving.

My drink smells or tastes like vinegar!

The most likely cause of vinegar flavors or aromas is exposure to a friendly bacteria called *Acetobactor*. This bacteria is most famous for helping make the vinegar we use for cooking, and it can wind up in your batch if fermentation was weak, sanitation wasn't quite perfect, or the batch was exposed to a lot of air at some point (like if you bottle using a spigot instead of a siphon).

If the flavor isn't overly unpleasant to you, it's fine to go ahead and drink it. If it tastes extremely vinegary, you can dilute it with sparkling water in the glass or toss the batch. You could also stash it in the back of the cupboard and wait for it to fully transform into vinegar, which you can then use for cooking.

Hard kombucha is especially vulnerable to vinegar flavors since exposure to air and friendly bacteria is part of the process for making it. You can keep things under control by having a strong, healthy scoby and using only kombucha that's 7 to 10 days old for making hard kombucha.

My drink smells or tastes like wet cardboard or old cooking sherry!

These flavors and aromas are a sign of oxidation, and you'll usually only find them in older bottles or bottles that weren't well sealed. Be sure to drink all your bottles within 3 months. If you don't mind the flavor, the bottles are fine to drink; otherwise toss the rest of the batch.

1

HARD
Seltzers

Hard seltzer is one of the easiest homemade alcoholic drinks you can make. All you have to do is mix together sugar, water, and yeast, and in about a week, you'll have a bona fide batch of mildly alcoholic (albeit flavorless) sparkling water. Now if that doesn't sound *terribly* exciting, that's OK: your next move is to give it flavor by adding fresh fruit, fruit juice, extracts, herbs, or spices.

Commercial hard seltzer brands typically flavor their products using extracts, concentrates, and natural flavors (flavor compounds derived from natural sources), which allows them to dial in their flavors very precisely and maintain consistency batch after batch. A few of the recipes in this chapter and those that follow use these flavoring ingredients, too—they're fun to play with, especially if the fruit itself is hard to find, such as passion fruit, or if it's a flavor that's tricky to capture, such as watermelon.

But as small-batch home seltzer-makers, we also have a host of other flavoring options at our disposal. This opens the door to flavor combinations and experiments that are impossible at a commercial scale. I'm talking about seltzers made with juicy summer plums (page 61), or a "gin and tonic" made with the traditional herbs and spices found in tonic water (page 72), or a boozy spin on crisp, refreshing spa water (page 67). Soon you'll be strolling the grocery store and wondering "Will it seltzer?" with everything you see.

What to Expect: When making homemade seltzers with fruit or fruit juice, the colors will be significantly darker than the crystal-clear or lightly tinted seltzers you buy off the shelf. Homemade seltzers are also typically less sweet than store-bought; check out the guide to sweetening on page 35 for crafting your perfect sip.

basic HARD SELTZER

AVERAGE ABV: 4.5 TO 5.5% ● **ABOUT 1 GALLON**

This recipe is the base for all the other hard seltzers in this chapter, and you can also use it to make your own unique creations. Don't skip the yeast nutrient— just as with humans, a diet of pure sugar isn't good for yeast, so giving it some nutrients helps the yeast stay healthy and happy. When you're ready to start experimenting, use the types and quantities of the flavorings from the other recipes as a guide.

 TRY THIS! Corn sugar isn't the only sugar you can use! Mix things up by using table sugar, brown sugar, honey, maple syrup, or any other fermentable sugar. Some of the flavor of the sugar will linger in your finished seltzer, so plan your additional flavorings accordingly.

12 ounces (2⅓ cups) corn sugar for fermentation, plus 1 ounce (3 tablespoons) for bottling

½ teaspoon champagne yeast

1 teaspoon yeast nutrient

Flavorings to taste

Simple syrup (optional)

day 1

Sanitize your fermenter, air lock, and whisk. In the fermenter, combine 1 gallon of water, 12 ounces of the corn sugar, and the champagne yeast. Whisk until the corn sugar is dissolved and the liquid is foamy on top, 30 to 60 seconds. Seal the fermenter, fill the air lock with sanitizer, and insert it into the fermenter.

Place the fermenter somewhere dark, slightly warm (70° to 80°F), and out of the way. You should start to see signs of fermentation (like bubbling in the air lock) within 24 to 48 hours.

day 2

In a small, sanitized measuring cup, dissolve ½ teaspoon of the yeast nutrient in 2 tablespoons of warm water. Add to the fermenter, reseal, and replace the air lock. Swirl gently to distribute and return to your fermentation spot.

day 3

In a small, sanitized measuring cup, dissolve ¼ teaspoon of the yeast nutrient in 2 tablespoons of warm water. Add to the fermenter, reseal, and replace the air lock. Swirl gently to distribute and return to your fermentation spot.

day 4

In a small, sanitized measuring cup, dissolve the remaining ¼ teaspoon yeast nutrient in 2 tablespoons of warm water. Add to the fermenter, reseal, and replace the air lock. Swirl gently to distribute and return to your fermentation spot.

days 5 to 14

Active fermentation will peak around Day 5 and then mostly finish around Day 7. Anytime between Day 7 and Day 10, add any flavorings to the fermenter and infuse for at least 3 days or up to 7 days.

After adding flavorings, you may see renewed signs of fermentation, though less vigorous than originally. Once you see no more signs of fermentation (like bubbles in the air lock), you can assume fermentation is complete. Wait another 24 hours to be safe, then proceed with bottling.

bottling day

When you're ready to bottle, sanitize a liquid measuring cup, spoon, large pot (1 gallon or larger), siphon, bottle filler, bottles, and caps. In the measuring cup, combine the remaining 1 ounce corn sugar with ½ cup of water and stir to dissolve. Pour this sugar water into the pot.

Siphon the seltzer into the pot with the sugar water, leaving behind any solids. Attach the bottle filler to your siphon, transfer the hard seltzer into the bottles, and cap.

Store somewhere cool, dark, and out of the way for 1 to 2 weeks to carbonate, or for up to 3 months. Chill before enjoying. For a sweeter drink, add a splash of simple syrup before serving.

lemon HARD SELTZER

AVERAGE ABV: 4.5 TO 5.5% **ABOUT 1 GALLON**

Easy peasy, lemon squeezy! This bright, citrusy hard seltzer is about as easy as it gets: easy to make, easy to sip, easy to love. This and the grapefruit seltzer (page 49) are my two "house seltzers" because they're nice to have on hand for a post-work sip or when friends stop by for pizza night.

TRY THIS! For a limoncello-esque flavor, add ½ cup of unsweetened coconut flakes along with the lemon. It's not enough to make this taste really coconutty, but it will add a soft sweetness.

12 ounces (2⅓ cups) corn sugar for fermentation, plus 1 ounce (3 tablespoons) for bottling

½ teaspoon champagne yeast

1 teaspoon yeast nutrient

Strips of zest and juice from 2 lemons

Lemon wedges, for garnish (optional)

Simple syrup (optional)

day 1

Sanitize your fermenter, air lock, and whisk. In the fermenter, combine 1 gallon of water, 12 ounces of the corn sugar, and the champagne yeast. Whisk until the corn sugar is dissolved and the liquid is foamy on top, 30 to 60 seconds. Seal the fermenter, fill the air lock with sanitizer, and insert it into the fermenter.

Place the fermenter somewhere dark, slightly warm (70° to 80°F), and out of the way. You should start to see signs of fermentation (like bubbling in the air lock) within 24 to 48 hours.

day 2

In a small, sanitized measuring cup, dissolve ½ teaspoon of the yeast nutrient in 2 tablespoons of warm water. Add to the fermenter, reseal, and replace the air lock. Swirl gently to distribute and return to your fermentation spot.

day 3

In a small, sanitized measuring cup, dissolve ¼ teaspoon of the yeast nutrient in 2 tablespoons of warm water. Add to the fermenter, reseal, and replace the air lock. Swirl gently to distribute and return to your fermentation spot.

day 4

In a small, sanitized measuring cup, dissolve the remaining ¼ teaspoon yeast nutrient in 2 tablespoons of warm water. Add to the fermenter, reseal, and replace the air lock. Swirl gently to distribute and return to your fermentation spot.

days 5 to 14

Active fermentation will peak around Day 5 and then mostly finish around Day 7. Anytime between Day 7 and Day 10, add the lemon zest and juice to the fermenter and infuse for at least 3 days or up to 7 days.

After adding flavorings, you may see renewed signs of fermentation, though less vigorous than originally. Once you see no more signs of fermentation (like bubbles in the air lock), you can assume fermentation is complete. Wait another 24 hours to be safe, then proceed with bottling.

bottling day

When you're ready to bottle, sanitize a liquid measuring cup, spoon, large pot (1 gallon or larger), siphon, bottle filler, bottles, and caps. In the measuring cup, combine the remaining 1 ounce corn sugar with ½ cup of water and stir to dissolve. Pour this sugar water into the pot.

Siphon the seltzer into the pot with the sugar water, leaving behind any solids. Attach the bottle filler to your siphon, transfer the hard seltzer into the bottles, and cap.

Store somewhere cool, dark, and out of the way for 1 to 2 weeks to carbonate, or for up to 3 months. Chill before enjoying and serve with garnish if desired. For a sweeter drink, add a splash of simple syrup before serving.

ruby red grapefruit
HARD SELTZER

AVERAGE ABV: 4.5 TO 5.5% ● **ABOUT 1 GALLON**

From my choice in candy to my choice in drinks, I've always been drawn to sour flavors more than sweet, so it was important for me to get this grapefruit seltzer right. I tried several kinds of fresh grapefruit as well as grapefruit flavoring and various combinations of all of these. In the end, the simplest was the best (as it so often is), and the balance of sweet and tart flavors from ruby red grapefruits was the winner for me.

 TRY THIS! For a sweeter grapefruit flavor, use 1 to 2 ounces of grapefruit flavoring (such as Brewer's Best) instead of the fresh grapefruit.

12 ounces (2⅓ cups) corn sugar for fermentation, plus 1 ounce (3 tablespoons) for bottling	Strips of zest and juice from 2 ruby red grapefruits
½ teaspoon champagne yeast	Fresh or dried grapefruit slices, for garnish (optional)
1 teaspoon yeast nutrient	Simple syrup (optional)

day 1	Sanitize your fermenter, air lock, and whisk. In the fermenter, combine 1 gallon of water, 12 ounces of the corn sugar, and the champagne yeast. Whisk until the corn sugar is dissolved and the liquid is foamy on top, 30 to 60 seconds. Seal the fermenter, fill the air lock with sanitizer, and insert it into the fermenter.
	Place the fermenter somewhere dark, slightly warm (70° to 80°F), and out of the way. You should start to see signs of fermentation (like bubbling in the air lock) within 24 to 48 hours.
day 2	In a small, sanitized measuring cup, dissolve ½ teaspoon of the yeast nutrient in 2 tablespoons of warm water. Add to the fermenter, reseal, and replace the air lock. Swirl gently to distribute and return to your fermentation spot.
day 3	In a small, sanitized measuring cup, dissolve ¼ teaspoon of the yeast nutrient in 2 tablespoons of warm water. Add to the fermenter, reseal, and replace the air lock. Swirl gently to distribute and return to your fermentation spot.

CONTINUED

day 4	In a small, sanitized measuring cup, dissolve the remaining ¼ teaspoon yeast nutrient in 2 tablespoons of warm water. Add to the fermenter, reseal, and replace the air lock. Swirl gently to distribute and return to your fermentation spot.
days 5 to 14	Active fermentation will peak around Day 5 and then mostly finish around Day 7. Anytime between Day 7 and Day 10, add the grapefruit zest and juice to the fermenter and infuse for at least 3 days or up to 7 days. After adding flavorings, you may see renewed signs of fermentation, though less vigorous than originally. Once you see no more signs of fermentation (like bubbles in the air lock), you can assume fermentation is complete. Wait another 24 hours to be safe, then proceed with bottling.
bottling day	When you're ready to bottle, sanitize a liquid measuring cup, spoon, large pot (1 gallon or larger), siphon, bottle filler, bottles, and caps. In the measuring cup, combine the remaining 1 ounce corn sugar with ½ cup of water and stir to dissolve. Pour this sugar water into the pot. Siphon the seltzer into the pot with the sugar water, leaving behind any solids. Attach the bottle filler to your siphon, transfer the hard seltzer into the bottles, and cap.

Store somewhere cool, dark, and out of the way for 1 to 2 weeks to carbonate, or for up to 3 months. Chill before enjoying and serve with garnish if desired. For a sweeter drink, add a splash of simple syrup before serving.

black cherry HARD SELTZER

AVERAGE ABV: 4.5 TO 5.5% ● **ABOUT 1 GALLON**

As a fruit, cherries don't get enough time in the spotlight, in my opinion, and this seltzer is pure cherry through and through. Using fresh or frozen cherries gives you real cherry flavor—there's no risk of this tasting like cough syrup, I promise! I call for sweet cherries here, but you could absolutely swap in tart cherries or a mix of sweet and tart cherries if you like. If you do, skip the lemon zest.

 TRY THIS! For a cherry-lime hard seltzer (yum!), skip the lemon zest and instead add strips of zest and juice from 2 limes.

12 ounces (2⅓ cups) corn sugar for fermentation, plus 1 ounce (3 tablespoons) for bottling

½ teaspoon champagne yeast

1 teaspoon yeast nutrient

1 pound (about 3¼ cups) fresh or frozen pitted sweet cherries, roughly chopped

Strips of zest from 1 lemon

Fresh pitted cherries, for garnish (optional)

Simple syrup (optional)

day 1	Sanitize your fermenter, air lock, and whisk. In the fermenter, combine 1 gallon of water, 12 ounces of the corn sugar, and the champagne yeast. Whisk until the corn sugar is dissolved and the liquid is foamy on top, 30 to 60 seconds. Seal the fermenter, fill the air lock with sanitizer, and insert it into the fermenter. Place the fermenter somewhere dark, slightly warm (70° to 80°F), and out of the way. You should start to see signs of fermentation (like bubbling in the air lock) within 24 to 48 hours.
day 2	In a small, sanitized measuring cup, dissolve ½ teaspoon of the yeast nutrient in 2 tablespoons of warm water. Add to the fermenter, reseal, and replace the air lock. Swirl gently to distribute and return to your fermentation spot.
day 3	In a small, sanitized measuring cup, dissolve ¼ teaspoon of the yeast nutrient in 2 tablespoons of warm water. Add to the fermenter, reseal, and replace the air lock. Swirl gently to distribute and return to your fermentation spot.
day 4	In a small, sanitized measuring cup, dissolve the remaining ¼ teaspoon yeast nutrient in 2 tablespoons of warm water. Add to the fermenter, reseal, and replace the air lock. Swirl gently to distribute and return to your fermentation spot.

CONTINUED

days 5 to 14

Active fermentation will peak around Day 5 and then mostly finish around Day 7. Anytime between Day 7 and Day 10, add the cherries and lemon zest to the fermenter and infuse for at least 3 days or up to 7 days.

After adding flavorings, you may see renewed signs of fermentation, though less vigorous than originally. Once you see no more signs of fermentation (like bubbles in the air lock), you can assume fermentation is complete. Wait another 24 hours to be safe, then proceed with bottling.

bottling day

When you're ready to bottle, sanitize a liquid measuring cup, spoon, large pot (1 gallon or larger), siphon, bottle filler, bottles, and caps. In the measuring cup, combine the remaining 1 ounce corn sugar with ½ cup of water and stir to dissolve. Pour this sugar water into the pot.

Siphon the seltzer into the pot with the sugar water, leaving behind any solids. Attach the bottle filler to your siphon, transfer the hard seltzer into the bottles, and cap.

Store somewhere cool, dark, and out of the way for 1 to 2 weeks to carbonate, or for up to 3 months. Chill before enjoying and serve with garnish if desired. For a sweeter drink, add a splash of simple syrup before serving.

coconut-lime HARD SELTZER

AVERAGE ABV: 4.5 TO 5.5% ● **ABOUT 1 GALLON**

Think of this as a lime seltzer with a kiss of coconut. The coconut is just enough to add a hint of tropical sweetness to the glass but not so much that it will knock you over. Look for the kind of coconut that comes in large flakes, not shredded coconut. I find it has a better flavor and is also easier to strain from the seltzer.

TRY THIS! If you would actually *like* to be knocked over with coconut, double the amount here. You might also think about adding ½ cup of reposado or añejo tequila for a paloma cocktail variation.

12 ounces (2⅓ cups) corn sugar for fermentation, plus 1 ounce (3 tablespoons) for bottling

½ teaspoon champagne yeast

1 teaspoon yeast nutrient

1⅓ ounces (¾ cup) unsweetened coconut flakes

Strips of zest and juice from 3 limes

Lime wedges, for garnish (optional)

Simple syrup (optional)

---

day 1

Sanitize your fermenter, air lock, and whisk. In the fermenter, combine 1 gallon of water, 12 ounces of the corn sugar, and the champagne yeast. Whisk until the corn sugar is dissolved and the liquid is foamy on top, 30 to 60 seconds. Seal the fermenter, fill the air lock with sanitizer, and insert it into the fermenter.

Place the fermenter somewhere dark, slightly warm (70° to 80°F), and out of the way. You should start to see signs of fermentation (like bubbling in the air lock) within 24 to 48 hours.

---

day 2

In a small, sanitized measuring cup, dissolve ½ teaspoon of the yeast nutrient in 2 tablespoons of warm water. Add to the fermenter, reseal, and replace the air lock. Swirl gently to distribute and return to your fermentation spot.

---

day 3

In a small, sanitized measuring cup, dissolve ¼ teaspoon of the yeast nutrient in 2 tablespoons of warm water. Add to the fermenter, reseal, and replace the air lock. Swirl gently to distribute and return to your fermentation spot.

---

day 4

In a small, sanitized measuring cup, dissolve the remaining ¼ teaspoon yeast nutrient in 2 tablespoons of warm water. Add to the fermenter, reseal, and replace the air lock. Swirl gently to distribute and return to your fermentation spot.

---

days
5 to 14

Active fermentation will peak around Day 5 and then mostly finish around Day 7. Anytime between Day 7 and Day 10, add the coconut flakes along with the lime zest and juice to the fermenter and infuse for at least 3 days or up to 7 days.

After adding flavorings, you may see renewed signs of fermentation, though less vigorous than originally. Once you see no more signs of fermentation (like bubbles in the air lock), you can assume fermentation is complete. Wait another 24 hours to be safe, then proceed with bottling.

bottling
day

When you're ready to bottle, sanitize a liquid measuring cup, spoon, large pot (1 gallon or larger), siphon, bottle filler, bottles, and caps. In the measuring cup, combine the remaining 1 ounce corn sugar with ½ cup of water and stir to dissolve. Pour this sugar water into the pot.

Siphon the seltzer into the pot with the sugar water, leaving behind any solids. Attach the bottle filler to your siphon, transfer the hard seltzer into the bottles, and cap.

Store somewhere cool, dark, and out of the way for 1 to 2 weeks to carbonate, or for up to 3 months. Chill before enjoying and serve with garnish if desired. For a sweeter drink, add a splash of simple syrup before serving.

watermelon-mint HARD SELTZER

AVERAGE ABV: 4.5 TO 5.5% ● **ABOUT 1 GALLON**

Watermelon is a surprisingly tricky flavor to replicate with fresh fruit, partly because watermelons can vary so much in their flavor intensity and also because so much of the fruit is literally water! I opt to use watermelon fruit flavoring over the fruit for seltzers. I recommend Brewer's Best; if you use another brand, just be sure to follow the package instructions. Regardless, add a small amount to start, taste, and then add more for a stronger flavor.

 TRY THIS! Watermelon and lime are also fantastic together. Try swapping the mint for strips of zest and juice from 2 limes.

12 ounces (2⅓ cups) corn sugar for fermentation, plus 1 ounce (3 tablespoons) for bottling

½ teaspoon champagne yeast

1 teaspoon yeast nutrient

¼ cup lightly packed, roughly torn mint leaves

1 to 2 ounces (2 to 4 tablespoons) watermelon fruit flavoring, such as Brewer's Best

Mint sprigs and fresh watermelon cubes, for garnish (optional)

Simple syrup (optional)

day 1	Sanitize your fermenter, air lock, and whisk. In the fermenter, combine 1 gallon of water, 12 ounces of the corn sugar, and the champagne yeast. Whisk until the corn sugar is dissolved and the liquid is foamy on top, 30 to 60 seconds. Seal the fermenter, fill the air lock with sanitizer, and insert it into the fermenter.
	Place the fermenter somewhere dark, slightly warm (70° to 80°F), and out of the way. You should start to see signs of fermentation (like bubbling in the air lock) within 24 to 48 hours.
day 2	In a small, sanitized measuring cup, dissolve ½ teaspoon of the yeast nutrient in 2 tablespoons of warm water. Add to the fermenter, reseal, and replace the air lock. Swirl gently to distribute and return to your fermentation spot.
day 3	In a small, sanitized measuring cup, dissolve ¼ teaspoon of the yeast nutrient in 2 tablespoons of warm water. Add to the fermenter, reseal, and replace the air lock. Swirl gently to distribute and return to your fermentation spot.

day 4	In a small, sanitized measuring cup, dissolve the remaining ¼ teaspoon yeast nutrient in 2 tablespoons of warm water. Add to the fermenter, reseal, and replace the air lock. Swirl gently to distribute and return to your fermentation spot.
days 5 to 14	Active fermentation will peak around Day 5 and then mostly finish around Day 7. Anytime between Day 7 and Day 10, add the mint to the fermenter and infuse for at least 3 days or up to 7 days. After adding flavorings, you may see renewed signs of fermentation, though less vigorous than originally. Once you see no more signs of fermentation (like bubbles in the air lock), you can assume fermentation is complete. Wait another 24 hours to be safe, then proceed with bottling.
bottling day	When you're ready to bottle, sanitize a liquid measuring cup, spoon, large pot (1 gallon or larger), siphon, bottle filler, bottles, and caps. In the measuring cup, combine the remaining 1 ounce corn sugar with ½ cup of water and stir to dissolve. Pour this sugar water along with 1 ounce of the watermelon fruit flavoring into the pot. Siphon the seltzer into the pot with the sugar water, leaving behind any solids. Using a sanitized wine thief or measuring cup, scoop out a little liquid and give it a taste. Add more watermelon flavoring until the desired flavor is reached. Attach the bottle filler to your siphon, transfer the hard seltzer into the bottles, and cap.
	Store somewhere cool, dark, and out of the way for 1 to 2 weeks to carbonate, or for up to 3 months. Chill before enjoying and serve with garnish if desired. For a sweeter drink, add a splash of simple syrup before serving.

pineapple HARD SELTZER

AVERAGE ABV: 4.5 TO 5.5% ● **ABOUT 1 GALLON**

This pineapple seltzer is about as close as you can get to sunshine in a bottle. Sunny in color and flavor, it's an easy-drinking seltzer that will absolutely make you smile no matter your mood. I like making this one in the summer for lazy porch hangouts and neighborhood backyard parties. Good vibes only!

 TRY THIS! Turn this into a piña colada seltzer by adding ¾ cup unsweetened coconut flakes along with the pineapple, plus ¼ cup golden rum at bottling.

12 ounces (2⅓ cups) corn sugar for fermentation, plus 1 ounce (3 tablespoons) for bottling

½ teaspoon champagne yeast

1 teaspoon yeast nutrient

1 pound (about 3½ cups) fresh or frozen pineapple chunks

Fresh pineapple wedges, for garnish (optional)

Simple syrup (optional)

---

day 1

Sanitize your fermenter, air lock, and whisk. In the fermenter, combine 1 gallon of water, 12 ounces of the corn sugar, and the champagne yeast. Whisk until the corn sugar is dissolved and the liquid is foamy on top, 30 to 60 seconds. Seal the fermenter, fill the air lock with sanitizer, and insert it into the fermenter.

Place the fermenter somewhere dark, slightly warm (70° to 80°F), and out of the way. You should start to see signs of fermentation (like bubbling in the air lock) within 24 to 48 hours.

---

day 2

In a small, sanitized measuring cup, dissolve ½ teaspoon of the yeast nutrient in 2 tablespoons of warm water. Add to the fermenter, reseal, and replace the air lock. Swirl gently to distribute and return to your fermentation spot.

---

day 3

In a small, sanitized measuring cup, dissolve ¼ teaspoon of the yeast nutrient in 2 tablespoons of warm water. Add to the fermenter, reseal, and replace the air lock. Swirl gently to distribute and return to your fermentation spot.

---

day 4

In a small, sanitized measuring cup, dissolve the remaining ¼ teaspoon yeast nutrient in 2 tablespoons of warm water. Add to the fermenter, reseal, and replace the air lock. Swirl gently to distribute and return to your fermentation spot.

CONTINUED

days 5 to 14

Active fermentation will peak around Day 5 and then mostly finish around Day 7. Anytime between Day 7 and Day 10, add the pineapple to the fermenter and infuse for at least 3 days or up to 7 days.

After adding flavorings, you may see renewed signs of fermentation, though less vigorous than originally. Once you see no more signs of fermentation (like bubbles in the air lock), you can assume fermentation is complete. Wait another 24 hours to be safe, then proceed with bottling.

bottling day

When you're ready to bottle, sanitize a liquid measuring cup, spoon, large pot (1 gallon or larger), siphon, bottle filler, bottles, and caps. In the measuring cup, combine the remaining 1 ounce corn sugar with ½ cup of water and stir to dissolve. Pour this sugar water into the pot.

Siphon the seltzer into the pot with the sugar water, leaving behind any solids. Attach the bottle filler to your siphon, transfer the hard seltzer into the bottles, and cap.

Store somewhere cool, dark, and out of the way for 1 to 2 weeks to carbonate, or for up to 3 months. Chill before enjoying and serve with garnish if desired. For a sweeter drink, add a splash of simple syrup before serving.

yuzu-plum HARD SELTZER

AVERAGE ABV: 4.5 TO 5.5% ● **ABOUT 1 GALLON**

Often overshadowed by the flashier summer fruits, plums don't get nearly the love they deserve. Pair them here with yuzu, an Asian citrus variety that tastes like a cross between sweet mandarin orange and bitter grapefruit, to make a sophisticated, pink-hued summer seltzer. Look for plums with deep-purple skins and sweet red flesh. You can find bottles of pressed yuzu juice at Asian markets or online.

 TRY THIS! Red and black plums are the easiest to find, but this seltzer can be made with any variety of plum or pluot available to you.

12 ounces (2⅓ cups) corn sugar for fermentation, plus 1 ounce (3 tablespoons) for bottling

½ teaspoon champagne yeast

1 teaspoon yeast nutrient

1 pound (6 to 8 medium) red or black plums, pitted and roughly chopped

1 to 2 ounces (2 to 4 tablespoons) bottled yuzu juice

Fresh plum slices, for garnish (optional)

Simple syrup (optional)

---

day 1

Sanitize your fermenter, air lock, and whisk. In the fermenter, combine 1 gallon of water, 12 ounces of the corn sugar, and the champagne yeast. Whisk until the corn sugar is dissolved and the liquid is foamy on top, 30 to 60 seconds. Seal the fermenter, fill the air lock with sanitizer, and insert it into the fermenter.

Place the fermenter somewhere dark, slightly warm (70° to 80°F), and out of the way. You should start to see signs of fermentation (like bubbling in the air lock) within 24 to 48 hours.

---

day 2

In a small, sanitized measuring cup, dissolve ½ teaspoon of the yeast nutrient in 2 tablespoons of warm water. Add to the fermenter, reseal, and replace the air lock. Swirl gently to distribute and return to your fermentation spot.

---

day 3

In a small, sanitized measuring cup, dissolve ¼ teaspoon of the yeast nutrient in 2 tablespoons of warm water. Add to the fermenter, reseal, and replace the air lock. Swirl gently to distribute and return to your fermentation spot.

CONTINUED

day 4	In a small, sanitized measuring cup, dissolve the remaining ¼ teaspoon yeast nutrient in 2 tablespoons of warm water. Add to the fermenter, reseal, and replace the air lock. Swirl gently to distribute and return to your fermentation spot.
days 5 to 14	Active fermentation will peak around Day 5 and then mostly finish around Day 7. Anytime between Day 7 and Day 10, add the plums to the fermenter and infuse for at least 3 days or up to 7 days. After adding flavorings, you may see renewed signs of fermentation, though less vigorous than originally. Once you see no more signs of fermentation (like bubbles in the air lock), you can assume fermentation is complete. Wait another 24 hours to be safe, then proceed with bottling.
bottling day	When you're ready to bottle, sanitize a liquid measuring cup, spoon, large pot (1 gallon or larger), siphon, bottle filler, bottles, and caps. In the measuring cup, combine the remaining 1 ounce corn sugar with ½ cup of water and stir to dissolve. Pour this sugar water along with 1 ounce of the yuzu juice into the pot. Siphon the seltzer into the pot with the sugar water, leaving behind any solids. Using a sanitized wine thief or measuring cup, scoop out a little liquid and give it a taste. Add more yuzu juice until the desired flavor is reached. Attach the bottle filler to your siphon, transfer the hard seltzer into the bottles, and cap.
	Store somewhere cool, dark, and out of the way for 1 to 2 weeks to carbonate, or for up to 3 months. Chill before enjoying and serve with garnish if desired. For a sweeter drink, add a splash of simple syrup before serving.

blackberry HARD SELTZER

AVERAGE ABV: 4.5 TO 5.5% ● **ABOUT 1 GALLON**

When I first started telling people that I was writing a book about hard seltzers, my friend Teddy begged me to include her favorite: "Please do a blackberry one. Pleeeeeease!" For someone like myself, who grew up in the Midwest picking blackberries off roadside bushes and popping them straight into my mouth, this was not a hard sell. I only hope I did Teddy proud.

 TRY THIS! For a lighter flavor, reduce the quantity of blackberries to ½ pound (about 2 cups). For a sweeter flavor, swap half the blackberries with blueberries.

12 ounces (2⅓ cups) corn sugar for fermentation, plus 1 ounce (3 tablespoons) for bottling

½ teaspoon champagne yeast

1 teaspoon yeast nutrient

1 pound (about 4 cups) fresh or frozen blackberries (lightly muddled if fresh)

Strips of zest from 1 lemon

Fresh blackberries, for garnish (optional)

Simple syrup (optional)

day 1	Sanitize your fermenter, air lock, and whisk. In the fermenter, combine 1 gallon of water, 12 ounces of the corn sugar, and the champagne yeast. Whisk until the corn sugar is dissolved and the liquid is foamy on top, 30 to 60 seconds. Seal the fermenter, fill the air lock with sanitizer, and insert it into the fermenter.
	Place the fermenter somewhere dark, slightly warm (70° to 80°F), and out of the way. You should start to see signs of fermentation (like bubbling in the air lock) within 24 to 48 hours.
day 2	In a small, sanitized measuring cup, dissolve ½ teaspoon of the yeast nutrient in 2 tablespoons of warm water. Add to the fermenter, reseal, and replace the air lock. Swirl gently to distribute and return to your fermentation spot.
day 3	In a small, sanitized measuring cup, dissolve ¼ teaspoon of the yeast nutrient in 2 tablespoons of warm water. Add to the fermenter, reseal, and replace the air lock. Swirl gently to distribute and return to your fermentation spot.
day 4	In a small, sanitized measuring cup, dissolve the remaining ¼ teaspoon yeast nutrient in 2 tablespoons of warm water. Add to the fermenter, reseal, and replace the air lock. Swirl gently to distribute and return to your fermentation spot.

**days
5 to 14**

Active fermentation will peak around Day 5 and then mostly finish around Day 7. Anytime between Day 7 and Day 10, add the blackberries and lemon zest to the fermenter and infuse for at least 3 days or up to 7 days.

After adding flavorings, you may see renewed signs of fermentation, though less vigorous than originally. Once you see no more signs of fermentation (like bubbles in the air lock), you can assume fermentation is complete. Wait another 24 hours to be safe, then proceed with bottling.

**bottling
day**

When you're ready to bottle, sanitize a liquid measuring cup, spoon, large pot (1 gallon or larger), siphon, bottle filler, bottles, and caps. In the measuring cup, combine the remaining 1 ounce corn sugar with ½ cup of water and stir to dissolve. Pour this sugar water into the pot.

Siphon the seltzer into the pot with the sugar water, leaving behind any solids. Attach the bottle filler to your siphon, transfer the hard seltzer into the bottles, and cap.

Store somewhere cool, dark, and out of the way for 1 to 2 weeks to carbonate, or for up to 3 months. Chill before enjoying and serve with garnish if desired. For a sweeter drink, add a splash of simple syrup before serving.

spa water HARD SELTZER

AVERAGE ABV: 4.5 TO 5.5% ● **ABOUT 1 GALLON**

Relaxing with a friend after massages recently, I joked that the only thing missing from the cucumber- and mint-infused water we were sipping was a shot of vodka. And thus, the idea for this spa water seltzer was born. There's no vodka, but this mildly boozy riff on spa water is sure to relax your shoulders and soothe your spirits without leaving you fuzzy-headed.

 TRY THIS! Add up to 2 cups of your favorite fresh fruit along with the cucumbers and mint! Strawberries, watermelon, lightly crushed blueberries, and cantaloupe are all really fun to try. A squeeze of lemon or lime juice is also a very nice complement.

12 ounces (2⅓ cups) corn sugar for fermentation, plus 1 ounce (3 tablespoons) for bottling

½ teaspoon champagne yeast

1 teaspoon yeast nutrient

1 English cucumber, sliced into thin rounds

½ cup loosely packed mint leaves

Fresh mint, for garnish (optional)

Simple syrup (optional)

---

day 1

Sanitize your fermenter, air lock, and whisk. In the fermenter, combine 1 gallon of water, 12 ounces of the corn sugar, and the champagne yeast. Whisk until the corn sugar is dissolved and the liquid is foamy on top, 30 to 60 seconds. Seal the fermenter, fill the air lock with sanitizer, and insert it into the fermenter.

Place the fermenter somewhere dark, slightly warm (70° to 80°F), and out of the way. You should start to see signs of fermentation (like bubbling in the air lock) within 24 to 48 hours.

---

day 2

In a small, sanitized measuring cup, dissolve ½ teaspoon of the yeast nutrient in 2 tablespoons of warm water. Add to the fermenter, reseal, and replace the air lock. Swirl gently to distribute and return to your fermentation spot.

---

day 3

In a small, sanitized measuring cup, dissolve ¼ teaspoon of the yeast nutrient in 2 tablespoons of warm water. Add to the fermenter, reseal, and replace the air lock. Swirl gently to distribute and return to your fermentation spot.

---

day 4

In a small, sanitized measuring cup, dissolve the remaining ¼ teaspoon yeast nutrient in 2 tablespoons of warm water. Add to the fermenter, reseal, and replace the air lock. Swirl gently to distribute and return to your fermentation spot.

CONTINUED

days 5 to 14	Active fermentation will peak around Day 5 and then mostly finish around Day 7. Anytime between Day 11 and Day 13, add the cucumber and mint to the fermenter and infuse for at least 1 day or up to 3 days. After adding flavorings, you may see renewed signs of fermentation, though less vigorous than originally. Once you see no more signs of fermentation (like bubbles in the air lock), you can assume fermentation is complete. Wait another 24 hours to be safe, then proceed with bottling.
bottling day	When you're ready to bottle, sanitize a liquid measuring cup, spoon, large pot (1 gallon or larger), siphon, bottle filler, bottles, and caps. In the measuring cup, combine the remaining 1 ounce corn sugar with ½ cup of water and stir to dissolve. Pour this sugar water into the pot. Siphon the seltzer into the pot with the sugar water, leaving behind any solids. Attach the bottle filler to your siphon, transfer the hard seltzer into the bottles, and cap.
	Store somewhere cool, dark, and out of the way for 1 to 2 weeks to carbonate, or for up to 3 months. Chill before enjoying and serve with garnish if desired. For a sweeter drink, add a splash of simple syrup before serving.

aperol spritz HARD SELTZER

AVERAGE ABV: 5 TO 6% ● **ABOUT 1 GALLON**

The Aperol spritz has enjoyed a small but dedicated fandom for many years, myself included, and it's easy to see why. This three-ingredient cocktail is a simple mix of Aperol (a bittersweet, citrus-forward Italian aperitivo), prosecco, and club soda, and it makes a light, refreshing predinner cocktail. This recipe swaps the prosecco and club soda for a seltzer base, putting the spotlight right where it should be: on the Aperol.

 TRY THIS! For a sweeter, fruitier seltzer, add a pound of fresh or frozen strawberries during fermentation. If bitter flavors are what you crave, swap the Aperol for its more intense cousin Campari.

9 ounces (1¾ cups) corn sugar for fermentation, plus 1 ounce (3 tablespoons) for bottling	1 (750ml) bottle Aperol
½ teaspoon champagne yeast	Strips of zest and juice from 1 orange
1 teaspoon yeast nutrient	Fresh orange slices, for garnish (optional)
	Simple syrup (optional)

day 1	Sanitize your fermenter, air lock, and whisk. In the fermenter, combine 3 quarts of water, 9 ounces of the corn sugar, and the champagne yeast. Whisk until the corn sugar is dissolved and the liquid is foamy on top, 30 to 60 seconds. Seal the fermenter, fill the air lock with sanitizer, and insert it into the fermenter.
	Place the fermenter somewhere dark, slightly warm (70° to 80°F), and out of the way. You should start to see signs of fermentation (like bubbling in the air lock) within 24 to 48 hours.
day 2	In a small, sanitized measuring cup, dissolve ½ teaspoon of the yeast nutrient in 2 tablespoons of warm water. Add to the fermenter, reseal, and replace the air lock. Swirl gently to distribute and return to your fermentation spot.
day 3	In a small, sanitized measuring cup, dissolve ¼ teaspoon of the yeast nutrient in 2 tablespoons of warm water. Add to the fermenter, reseal, and replace the air lock. Swirl gently to distribute and return to your fermentation spot.

CONTINUED

day 4	In a small, sanitized measuring cup, dissolve the remaining ¼ teaspoon yeast nutrient in 2 tablespoons of warm water. Add to the fermenter, reseal, and replace the air lock. Swirl gently to distribute and return to your fermentation spot.
days 5 to 14	Active fermentation will peak around Day 5 and then mostly finish around Day 7. Anytime between Day 5 and Day 7, add the Aperol along with the orange zest and juice to the fermenter. After adding flavorings, you will see renewed signs of fermentation. Wait another 1 to 2 weeks, until you see no more signs of fermentation (like bubbles in the air lock), and then proceed with bottling.
bottling day	When you're ready to bottle, sanitize a liquid measuring cup, spoon, large pot (1 gallon or larger), siphon, bottle filler, bottles, and caps. In the measuring cup, combine the remaining 1 ounce corn sugar with ½ cup of water and stir to dissolve. Pour this sugar water into the pot. Siphon the seltzer into the pot with the sugar water, leaving behind any solids. Attach the bottle filler to your siphon, transfer the hard seltzer into the bottles, and cap.
	Store somewhere cool, dark, and out of the way for 1 to 2 weeks to carbonate, or for up to 3 months. Chill before enjoying and serve with garnish if desired. For a sweeter drink, add a splash of simple syrup before serving. (This drink becomes more clear and less red as it ages, though the flavor remains the same. Add a teaspoon or two of Aperol to the glass if you'd like a stronger red tint in your drink.)

gin & tonic HARD SELTZER

AVERAGE ABV: 4.5 TO 5.5% ● **ABOUT 1 GALLON**

The primary ingredient that makes tonic water taste like, well, tonic water is bark from the South American cinchona tree. On its own, cinchona is intensely bitter and not very palatable. But partner it with a handful of spices and citrus zest, give it some fizz, and you've got yourself a drink you can't stop sipping. Cinchona bark is most easily found online; look for 100% natural red cinchona bark. This hard seltzer tends to finish fairly dry, so I highly recommend some simple syrup in your glass when serving.

 TRY THIS! Make your tonic water more herbaceous with a few sprigs of rosemary and thyme. You can also up the amount of dried berries and peppercorns for a spicier kick.

12 ounces (2⅓ cups) corn sugar for fermentation, plus 1 ounce (3 tablespoons) for bottling

½ teaspoon champagne yeast

1 teaspoon yeast nutrient

Strips of zest from 1 lemon

Strips of zest from 1 lime

Strips of zest from 1 grapefruit

½ stalk lemongrass, sliced into thin rounds (optional)

1 tablespoon cinchona bark

5 whole allspice berries

¼ teaspoon black peppercorns

2 tablespoons juniper berries

1 teaspoon acid blend (see page 14)

Fresh lime wedges, for garnish (optional)

Simple syrup (optional)

day 1

Sanitize your fermenter, air lock, and whisk. In the fermenter, combine 1 gallon of water, 12 ounces of the corn sugar, and the champagne yeast. Whisk until the corn sugar is dissolved and the liquid is foamy on top, 30 to 60 seconds. Seal the fermenter, fill the air lock with sanitizer, and insert it into the fermenter.

Place the fermenter somewhere dark, slightly warm (70° to 80°F), and out of the way. You should start to see signs of fermentation (like bubbling in the air lock) within 24 to 48 hours.

day 2

In a small, sanitized measuring cup, dissolve ½ teaspoon of the yeast nutrient in 2 tablespoons of warm water. Add to the fermenter, reseal, and replace the air lock. Swirl gently to distribute and return to your fermentation spot.

day 3	In a small, sanitized measuring cup, dissolve ¼ teaspoon of the yeast nutrient in 2 tablespoons of warm water. Add to the fermenter, reseal, and replace the air lock. Swirl gently to distribute and return to your fermentation spot.
day 4	In a small, sanitized measuring cup, dissolve the remaining ¼ teaspoon yeast nutrient in 2 tablespoons of warm water. Add to the fermenter, reseal, and replace the air lock. Swirl gently to distribute and return to your fermentation spot.
days 5 to 14	Active fermentation will peak around Day 5 and then mostly finish around Day 7. Anytime between Day 7 and Day 10, add the citrus zest, lemongrass, cinchona bark, allspice berries, peppercorns, juniper berries, and acid blend to the fermenter and infuse for at least 3 days or up to 7 days.
	After 3 days of infusing, use a sanitized wine thief or measuring cup to scoop out a little liquid and give it a taste. It should taste like flat tonic water—bitter from the cinchona bark, botanical from the zest and spices, and tart from the acid blend. Add more of any of the flavorings if desired and infuse for another 1 to 3 days.
	After adding flavorings, you may see renewed signs of fermentation, though less vigorous than originally. Once you see no more signs of fermentation (like bubbles in the air lock), you can assume fermentation is complete. Wait another 24 hours to be safe, then proceed with bottling.
bottling day	When you're ready to bottle, sanitize a liquid measuring cup, spoon, large pot (1 gallon or larger), siphon, bottle filler, bottles, and caps. In the measuring cup, combine the remaining 1 ounce corn sugar with ½ cup of water and stir to dissolve. Pour this sugar water into the pot.
	Siphon the seltzer into the pot with the sugar water, leaving behind any solids. Attach the bottle filler to your siphon, transfer the hard seltzer into the bottles, and cap.

Store somewhere cool, dark, and out of the way for 1 to 2 weeks to carbonate, or for up to 3 months. Chill before enjoying and serve with garnish if desired. For a sweeter drink, add a splash of simple syrup before serving.

2

HARD
Iced Teas

When you're ready for happy hour but also feeling a bit sleepy, hard iced teas are here for you. Not only are they delicious, but they also deliver a little hit of caffeine. You won't get quite the same caffeine buzz as a strong cup of tea (fermentation breaks down some of the natural caffeine), but enough lingers to give you a boost.

The process for making hard iced tea isn't very different from making hard seltzer. In fact, they start the same way: mix together some sugar and water, add yeast, and let it ferment. The iced tea is introduced during the flavoring step once the bulk of fermentation has ended. This gives us a cleaner, purer tea flavor than if we added it at the start of fermentation. Using artesian water instead of tap water also helps us achieve a better tea flavor.

In some of these recipes, like the Raspberry Hard Iced Tea (page 79) and the Peach Hard Iced Tea (page 81), the iced tea is the backbone for the drink and a strong flavor component. In others, like the Sangria Hard Iced Tea (page 101) and the Mai Tai Hard Iced Tea (page 103), it's more of a backdrop for other ingredients and flavors—it's not the star, but it helps the stars shine.

What to Expect: Like the other recipes in this book, homemade hard iced teas will be less sweet than store-bought versions you may have tried in the past. A *lot* less, since commercial hard iced teas tend to be heavy on the sweeteners. If you find yourself craving that sweetness, I recommend adding some simple syrup to the glass when serving.

basic HARD ICED TEA

AVERAGE ABV: 4.5 TO 5.5% ● **ABOUT 1 GALLON**

Making hard iced tea is as easy as making basic hard seltzer (page 44) and throwing in a few tea bags, with one exception: *the water you use for hard iced tea really matters.* My first few batches of hard iced tea were weak tasting and muddy flavored, and then I finally realized that my hard tap water was preventing the tea from fully infusing. Once I made the switch to store-bought spring water, the clouds parted, the sun came out, and hard iced tea was mine. Even if your water isn't particularly hard (lucky you!), I still recommend using spring, artesian, or reverse osmosis water for the best flavor with these recipes.

⬤ **TRY THIS!** Want the flavor of iced tea but not the caffeine? Use decaffeinated tea instead!

1 gallon spring, artesian, or reverse osmosis water

12 ounces (2⅓ cups) corn sugar for fermentation, plus 1 ounce (3 tablespoons) for bottling

½ teaspoon champagne yeast

1 teaspoon yeast nutrient

10 bags good-quality black tea

Flavorings to taste

Simple syrup (optional)

---

day 1

Sanitize your fermenter, air lock, and whisk. In the fermenter, combine the water, 12 ounces of the corn sugar, and the champagne yeast. Whisk until the corn sugar is dissolved and the liquid is foamy on top, 30 to 60 seconds. Seal the fermenter, fill the air lock with sanitizer, and insert it into the fermenter.

Place the fermenter somewhere dark, slightly warm (70° to 80°F), and out of the way. You should start to see signs of fermentation (like bubbling in the air lock) within 24 to 48 hours.

---

day 2

In a small, sanitized measuring cup, dissolve ½ teaspoon of the yeast nutrient in 2 tablespoons of warm water. Add to the fermenter, reseal, and replace the air lock. Swirl gently to distribute and return to your fermentation spot.

---

day 3

In a small, sanitized measuring cup, dissolve ¼ teaspoon of the yeast nutrient in 2 tablespoons of warm water. Add to the fermenter, reseal, and replace the air lock. Swirl gently to distribute and return to your fermentation spot.

---

day 4	In a small, sanitized measuring cup, dissolve the remaining ¼ teaspoon yeast nutrient in 2 tablespoons of warm water. Add to the fermenter, reseal, and replace the air lock. Swirl gently to distribute and return to your fermentation spot.
days 5 to 14	Active fermentation will peak around Day 5 and then mostly finish around Day 7. Anytime between Day 7 and Day 10, add the tea along with any flavorings to the fermenter and infuse for at least 3 days or up to 7 days. After adding flavorings, you may see renewed signs of fermentation, though less vigorous than originally. Once you see no more signs of fermentation (like bubbles in the air lock), you can assume fermentation is complete. Wait another 24 hours to be safe, then proceed with bottling.
bottling day	When you're ready to bottle, sanitize a liquid measuring cup, spoon, large pot (1 gallon or larger), siphon, bottle filler, bottles, and caps. In the measuring cup, combine the remaining 1 ounce corn sugar with ½ cup of water and stir to dissolve. Pour this sugar water into the pot. Siphon the hard iced tea into the pot with the sugar water, leaving behind the tea bags and any solids. Attach the bottle filler to your siphon, transfer the hard iced tea into the bottles, and cap.
	Store somewhere cool, dark, and out of the way for 1 to 2 weeks to carbonate, or for up to 3 months. Chill before enjoying. For a sweeter drink, add a splash of simple syrup before serving.

raspberry HARD ICED TEA

AVERAGE ABV: 4.5 TO 5.5% ● **ABOUT 1 GALLON**

Take your hard iced tea base, add enough tart red raspberries to make things interesting, and you've got yourself one lip-smacking brew in a bottle. This is the kind of beverage that I like to sip on summer afternoons while I'm outside grilling. It perks me up without making me too fuzzy-headed; plus, it goes great with whatever is coming off the grill.

 TRY THIS! Add ¼ cup lightly packed mint or basil leaves along with the raspberries for another layer of goodness.

1 gallon spring, artesian, or reverse osmosis water

12 ounces (2⅓ cups) corn sugar for fermentation, plus 1 ounce (3 tablespoons) for bottling

½ teaspoon champagne yeast

1 teaspoon yeast nutrient

10 bags good-quality black tea

12 ounces (about 3 cups) fresh or frozen raspberries (lightly muddled if fresh)

Fresh raspberries, for garnish (optional)

Simple syrup (optional)

---

day 1

Sanitize your fermenter, air lock, and whisk. In the fermenter, combine the water, 12 ounces of the corn sugar, and the champagne yeast. Whisk until the corn sugar is dissolved and the liquid is foamy on top, 30 to 60 seconds. Seal the fermenter, fill the air lock with sanitizer, and insert it into the fermenter.

Place the fermenter somewhere dark, slightly warm (70° to 80°F), and out of the way. You should start to see signs of fermentation (like bubbling in the air lock) within 24 to 48 hours.

---

day 2

In a small, sanitized measuring cup, dissolve ½ teaspoon of the yeast nutrient in 2 tablespoons of warm water. Add to the fermenter, reseal, and replace the air lock. Swirl gently to distribute and return to your fermentation spot.

---

day 3

In a small, sanitized measuring cup, dissolve ¼ teaspoon of the yeast nutrient in 2 tablespoons of warm water. Add to the fermenter, reseal, and replace the air lock. Swirl gently to distribute and return to your fermentation spot.

CONTINUED

day 4	In a small, sanitized measuring cup, dissolve the remaining ¼ teaspoon yeast nutrient in 2 tablespoons of warm water. Add to the fermenter, reseal, and replace the air lock. Swirl gently to distribute and return to your fermentation spot.
days 5 to 14	Active fermentation will peak around Day 5 and then mostly finish around Day 7. Anytime between Day 7 and Day 10, add the tea along with the raspberries to the fermenter and infuse for at least 3 days or up to 7 days.
	After adding flavorings, you may see renewed signs of fermentation, though less vigorous than originally. Once you see no more signs of fermentation (like bubbles in the air lock), you can assume fermentation is complete. Wait another 24 hours to be safe, then proceed with bottling.
bottling day	When you're ready to bottle, sanitize a liquid measuring cup, spoon, large pot (1 gallon or larger), siphon, bottle filler, bottles, and caps. In the measuring cup, combine the remaining 1 ounce corn sugar with ½ cup of water and stir to dissolve. Pour this sugar water into the pot.
	Siphon the hard iced tea into the pot with the sugar water, leaving behind the tea bags and any solids. Attach the bottle filler to your siphon, transfer the hard iced tea into the bottles, and cap.
	Store somewhere cool, dark, and out of the way for 1 to 2 weeks to carbonate, or for up to 3 months. Chill before enjoying and serve with garnish if desired. For a sweeter drink, add a splash of simple syrup before serving.

peach HARD ICED TEA

AVERAGE ABV: 4.5 TO 5.5%   ● **ABOUT 1 GALLON**

The sweetness of summer peaches set against the soft earthiness of black tea is such a winning combo for me. It feels both playful and sophisticated, though that might have to do with the summer between my sophomore and junior years of high school, when I decided that bottled peach iced tea was superior to soda and would be key to my identity as an upperclassman. Regardless, it's an easy pairing to love, then and now.

TRY THIS! I wouldn't be mad at a splash of bourbon in my peach iced tea. Try soaking ½ ounce of oak cubes in ½ cup of bourbon for about a week, and then adding both of them along with the peaches.

1 gallon spring, artesian, or reverse osmosis water

12 ounces (2⅓ cups) corn sugar for fermentation, plus 1 ounce (3 tablespoons) for bottling

½ teaspoon champagne yeast

1 teaspoon yeast nutrient

10 bags good-quality black tea

½ pound (2 to 3 medium) fresh or frozen sliced peaches, peeled and pitted

Strips of zest from 1 lemon

Peach slices, for garnish (optional)

Simple syrup (optional)

---

day 1

Sanitize your fermenter, air lock, and whisk. In the fermenter, combine the water, 12 ounces of the corn sugar, and the champagne yeast. Whisk until the corn sugar is dissolved and the liquid is foamy on top, 30 to 60 seconds. Seal the fermenter, fill the air lock with sanitizer, and insert it into the fermenter.

Place the fermenter somewhere dark, slightly warm (70° to 80°F), and out of the way. You should start to see signs of fermentation (like bubbling in the air lock) within 24 to 48 hours.

---

day 2

In a small, sanitized measuring cup, dissolve ½ teaspoon of the yeast nutrient in 2 tablespoons of warm water. Add to the fermenter, reseal, and replace the air lock. Swirl gently to distribute and return to your fermentation spot.

---

day 3

In a small, sanitized measuring cup, dissolve ¼ teaspoon of the yeast nutrient in 2 tablespoons of warm water. Add to the fermenter, reseal, and replace the air lock. Swirl gently to distribute and return to your fermentation spot.

CONTINUED

day 4	In a small, sanitized measuring cup, dissolve the remaining ¼ teaspoon yeast nutrient in 2 tablespoons of warm water. Add to the fermenter, reseal, and replace the air lock. Swirl gently to distribute and return to your fermentation spot.
days 5 to 14	Active fermentation will peak around Day 5 and then mostly finish around Day 7. Anytime between Day 7 and Day 10, add the tea along with the peaches and lemon zest to the fermenter and infuse for at least 3 days or up to 7 days. After adding flavorings, you may see renewed signs of fermentation, though less vigorous than originally. Once you see no more signs of fermentation (like bubbles in the air lock), you can assume fermentation is complete. Wait another 24 hours to be safe, then proceed with bottling.
bottling day	When you're ready to bottle, sanitize a liquid measuring cup, spoon, large pot (1 gallon or larger), siphon, bottle filler, bottles, and caps. In the measuring cup, combine the remaining 1 ounce corn sugar with ½ cup of water and stir to dissolve. Pour this sugar water into the pot. Siphon the hard iced tea into the pot with the sugar water, leaving behind the tea bags and any solids. Attach the bottle filler to your siphon, transfer the hard iced tea into the bottles, and cap.

Store somewhere cool, dark, and out of the way for 1 to 2 weeks to carbonate, or for up to 3 months. Chill before enjoying and serve with garnish if desired. For a sweeter drink, add a splash of simple syrup before serving.

summer melon HARD ICED TEA

AVERAGE ABV: 4.5 TO 5.5% ● **ABOUT 1 GALLON**

This recipe requires cantaloupe at the very peak of the summer season. I'm talking about the kind that leaves juice all over the counter after you cut into it. It should taste like golden honey and be so ripe that it melts in your mouth. So go ahead—thump every melon you can reach until you find the perfect cantaloupe in the bunch.

 TRY THIS! A few sprigs of mint along with the cantaloupe would be fantastic. You can also swap in honeydew or any other summer melon.

1 gallon spring, artesian, or reverse osmosis water

12 ounces (2⅓ cups) corn sugar for fermentation, plus 1 ounce (3 tablespoons) for bottling

½ teaspoon champagne yeast

1 teaspoon yeast nutrient

10 bags good-quality black tea

½ pound (about 2 cups) cantaloupe cubes, from about ½ melon

Melon cubes, for garnish (optional)

Simple syrup (optional)

day 1	Sanitize your fermenter, air lock, and whisk. In the fermenter, combine the water, 12 ounces of the corn sugar, and the champagne yeast. Whisk until the corn sugar is dissolved and the liquid is foamy on top, 30 to 60 seconds. Seal the fermenter, fill the air lock with sanitizer, and insert it into the fermenter.
	Place the fermenter somewhere dark, slightly warm (70° to 80°F), and out of the way. You should start to see signs of fermentation (like bubbling in the air lock) within 24 to 48 hours.
day 2	In a small, sanitized measuring cup, dissolve ½ teaspoon of the yeast nutrient in 2 tablespoons of warm water. Add to the fermenter, reseal, and replace the air lock. Swirl gently to distribute and return to your fermentation spot.
day 3	In a small, sanitized measuring cup, dissolve ¼ teaspoon of the yeast nutrient in 2 tablespoons of warm water. Add to the fermenter, reseal, and replace the air lock. Swirl gently to distribute and return to your fermentation spot.

day 4	In a small, sanitized measuring cup, dissolve the remaining ¼ teaspoon yeast nutrient in 2 tablespoons of warm water. Add to the fermenter, reseal, and replace the air lock. Swirl gently to distribute and return to your fermentation spot.
days 5 to 14	Active fermentation will peak around Day 5 and then mostly finish around Day 7. Anytime between Day 7 and Day 10, add the tea along with the cantaloupe to the fermenter and infuse for at least 3 days or up to 7 days. After adding flavorings, you may see renewed signs of fermentation, though less vigorous than originally. Once you see no more signs of fermentation (like bubbles in the air lock), you can assume fermentation is complete. Wait another 24 hours to be safe, then proceed with bottling.
bottling day	When you're ready to bottle, sanitize a liquid measuring cup, spoon, large pot (1 gallon or larger), siphon, bottle filler, bottles, and caps. In the measuring cup, combine the remaining 1 ounce corn sugar with ½ cup of water and stir to dissolve. Pour this sugar water into the pot. Siphon the hard iced tea into the pot with the sugar water, leaving behind the tea bags and any solids. Attach the bottle filler to your siphon, transfer the hard iced tea into the bottles, and cap.

Store somewhere cool, dark, and out of the way for 1 to 2 weeks to carbonate, or for up to 3 months. Chill before enjoying and serve with garnish if desired. For a sweeter drink, add a splash of simple syrup before serving.

agua de jamaica HARD ICED TEA

AVERAGE ABV: 4.5 TO 5.5% ● **ABOUT 1 GALLON**

Tangy, fruity agua de jamaica is a standard at Mexican restaurants and taco trucks everywhere. It's made by steeping dried hibiscus flowers and is often further flavored with cinnamon, citrus, fresh ginger, or all of the above. Make sure to buy true hibiscus flowers for this recipe, not bags of tea or anything already mixed with other ingredients. You can find hibiscus flowers, or flor de jamaica, at Mexican groceries, loose-leaf tea vendors, or online.

 TRY THIS! Mint is a great addition to agua de jamaica. Try adding ¼ cup of loosely packed leaves along with the other flavoring ingredients.

1 gallon spring, artesian, or reverse osmosis water

12 ounces (2⅓ cups) corn sugar for fermentation, plus 1 ounce (3 tablespoons) for bottling

½ teaspoon champagne yeast

1 teaspoon yeast nutrient

10 bags good-quality black tea

⅓ ounce (½ cup) dried hibiscus flowers

1 cinnamon stick

Strips of zest and juice from 1 lime

½ ounce (2 tablespoons) roughly chopped ginger

Lime wedges, for garnish (optional)

Simple syrup (optional)

---

day 1

Sanitize your fermenter, air lock, and whisk. In the fermenter, combine the water, 12 ounces of the corn sugar, and the champagne yeast. Whisk until the corn sugar is dissolved and the liquid is foamy on top, 30 to 60 seconds. Seal the fermenter, fill the air lock with sanitizer, and insert it into the fermenter.

Place the fermenter somewhere dark, slightly warm (70° to 80°F), and out of the way. You should start to see signs of fermentation (like bubbling in the air lock) within 24 to 48 hours.

---

day 2

In a small, sanitized measuring cup, dissolve ½ teaspoon of the yeast nutrient in 2 tablespoons of warm water. Add to the fermenter, reseal, and replace the air lock. Swirl gently to distribute and return to your fermentation spot.

---

day 3

In a small, sanitized measuring cup, dissolve ¼ teaspoon of the yeast nutrient in 2 tablespoons of warm water. Add to the fermenter, reseal, and replace the air lock. Swirl gently to distribute and return to your fermentation spot.

CONTINUED

day 4	In a small, sanitized measuring cup, dissolve the remaining ¼ teaspoon yeast nutrient in 2 tablespoons of warm water. Add to the fermenter, reseal, and replace the air lock. Swirl gently to distribute and return to your fermentation spot.
days 5 to 14	Active fermentation will peak around Day 5 and then mostly finish around Day 7. Anytime between Day 7 and Day 10, add the tea along with the hibiscus, cinnamon, lime zest and juice, and ginger to the fermenter and infuse for at least 3 days or up to 7 days.
	After adding flavorings, you may see renewed signs of fermentation, though less vigorous than originally. Once you see no more signs of fermentation (like bubbles in the air lock), you can assume fermentation is complete. Wait another 24 hours to be safe, then proceed with bottling.
bottling day	When you're ready to bottle, sanitize a liquid measuring cup, spoon, large pot (1 gallon or larger), siphon, bottle filler, bottles, and caps. In the measuring cup, combine the remaining 1 ounce corn sugar with ½ cup of water and stir to dissolve. Pour this sugar water into the pot.
	Siphon the hard iced tea into the pot with the sugar water, leaving behind the tea bags and any solids. Attach the bottle filler to your siphon, transfer the hard iced tea into the bottles, and cap.
	Store somewhere cool, dark, and out of the way for 1 to 2 weeks to carbonate, or for up to 3 months. Chill before enjoying and serve with garnish if desired. For a sweeter drink, add a splash of simple syrup before serving.

blueberry jasmine HARD ICED TEA

AVERAGE ABV: 4.5 TO 5.5% ● **ABOUT 1 GALLON**

Jasmine and blueberries both have such delicate aromas and flavors that they can easily be overwhelmed when other, bolder ingredients are present. So it makes sense that they are each other's best friend. Blueberries lend their fruity sweetness to this sparkling drink, while jasmine green tea brings soft floral notes—neither overshadowing the other, both helping the other shine.

 TRY THIS! Instead of jasmine tea, try Earl Grey. The citrusy bergamot flavor in this tea would be excellent with blueberries.

1 gallon spring, artesian, or reverse osmosis water

12 ounces (2⅓ cups) corn sugar for fermentation, plus 1 ounce (3 tablespoons) for bottling

½ teaspoon champagne yeast

1 teaspoon yeast nutrient

10 bags good-quality jasmine green tea

½ pound (about 2 scant cups) fresh or frozen blueberries (lightly muddled if fresh)

Blueberries, for garnish (optional)

Simple syrup (optional)

---

day 1

Sanitize your fermenter, air lock, and whisk. In the fermenter, combine the water, 12 ounces of the corn sugar, and the champagne yeast. Whisk until the corn sugar is dissolved and the liquid is foamy on top, 30 to 60 seconds. Seal the fermenter, fill the air lock with sanitizer, and insert it into the fermenter.

Place the fermenter somewhere dark, slightly warm (70° to 80°F), and out of the way. You should start to see signs of fermentation (like bubbling in the air lock) within 24 to 48 hours.

---

day 2

In a small, sanitized measuring cup, dissolve ½ teaspoon of the yeast nutrient in 2 tablespoons of warm water. Add to the fermenter, reseal, and replace the air lock. Swirl gently to distribute and return to your fermentation spot.

---

day 3

In a small, sanitized measuring cup, dissolve ¼ teaspoon of the yeast nutrient in 2 tablespoons of warm water. Add to the fermenter, reseal, and replace the air lock. Swirl gently to distribute and return to your fermentation spot.

CONTINUED

day 4	In a small, sanitized measuring cup, dissolve the remaining ¼ teaspoon yeast nutrient in 2 tablespoons of warm water. Add to the fermenter, reseal, and replace the air lock. Swirl gently to distribute and return to your fermentation spot.
days 5 to 14	Active fermentation will peak around Day 5 and then mostly finish around Day 7. Anytime between Day 7 and Day 10, add the tea along with the blueberries to the fermenter and infuse for at least 3 days or up to 7 days. After adding flavorings, you may see renewed signs of fermentation, though less vigorous than originally. Once you see no more signs of fermentation (like bubbles in the air lock), you can assume fermentation is complete. Wait another 24 hours to be safe, then proceed with bottling.
bottling day	When you're ready to bottle, sanitize a liquid measuring cup, spoon, large pot (1 gallon or larger), siphon, bottle filler, bottles, and caps. In the measuring cup, combine the remaining 1 ounce corn sugar with ½ cup of water and stir to dissolve. Pour this sugar water into the pot. Siphon the hard iced tea into the pot with the sugar water, leaving behind the tea bags and any solids. Attach the bottle filler to your siphon, transfer the hard iced tea into the bottles, and cap.
	Store somewhere cool, dark, and out of the way for 1 to 2 weeks to carbonate, or for up to 3 months. Chill before enjoying and serve with garnish if desired. For a sweeter drink, add a splash of simple syrup before serving.

hard ARNOLD PALMER

AVERAGE ABV: 4.5 TO 5.5% ● **ABOUT 1 GALLON**

In my humble opinion, there is nothing more refreshing than an Arnold Palmer on a hot summer day. The combo of ice-cold tea plus fresh-squeezed lemonade is just unbeatable. Now give it some fizz and add a touch of alcohol, and we've got ourselves the perfect grown-up lawn chair drink, don't you think?

 TRY THIS! Swap the black tea for green tea and use honey instead of corn sugar for a sweet spin on this classic flavor combo.

1 gallon spring, artesian, or reverse osmosis water

12 ounces (2⅓ cups) corn sugar for fermentation, plus 1 ounce (3 tablespoons) for bottling

½ teaspoon champagne yeast

1 teaspoon yeast nutrient

10 bags good-quality black tea

Strips of zest and juice from 2 lemons

Lemon wedges, for garnish (optional)

Simple syrup (optional)

day 1	Sanitize your fermenter, air lock, and whisk. In the fermenter, combine the water, 12 ounces of the corn sugar, and the champagne yeast. Whisk until the corn sugar is dissolved and the liquid is foamy on top, 30 to 60 seconds. Seal the fermenter, fill the air lock with sanitizer, and insert it into the fermenter.
	Place the fermenter somewhere dark, slightly warm (70° to 80°F), and out of the way. You should start to see signs of fermentation (like bubbling in the air lock) within 24 to 48 hours.
day 2	In a small, sanitized measuring cup, dissolve ½ teaspoon of the yeast nutrient in 2 tablespoons of warm water. Add to the fermenter, reseal, and replace the air lock. Swirl gently to distribute and return to your fermentation spot.
day 3	In a small, sanitized measuring cup, dissolve ¼ teaspoon of the yeast nutrient in 2 tablespoons of warm water. Add to the fermenter, reseal, and replace the air lock. Swirl gently to distribute and return to your fermentation spot.
day 4	In a small, sanitized measuring cup, dissolve the remaining ¼ teaspoon yeast nutrient in 2 tablespoons of warm water. Add to the fermenter, reseal, and replace the air lock. Swirl gently to distribute and return to your fermentation spot.

CONTINUED

days 5 to 14

Active fermentation will peak around Day 5 and then mostly finish around Day 7. Anytime between Day 7 and Day 10, add the tea along with the lemon zest and juice to the fermenter and infuse for at least 3 days or up to 7 days.

After adding flavorings, you may see renewed signs of fermentation, though less vigorous than originally. Once you see no more signs of fermentation (like bubbles in the air lock), you can assume fermentation is complete. Wait another 24 hours to be safe, then proceed with bottling.

bottling day

When you're ready to bottle, sanitize a liquid measuring cup, spoon, large pot (1 gallon or larger), siphon, bottle filler, bottles, and caps. In the measuring cup, combine the remaining 1 ounce corn sugar with ½ cup of water and stir to dissolve. Pour this sugar water into the pot.

Siphon the hard iced tea into the pot with the sugar water, leaving behind the tea bags and any solids. Attach the bottle filler to your siphon, transfer the hard iced tea into the bottles, and cap.

Store somewhere cool, dark, and out of the way for 1 to 2 weeks to carbonate, or for up to 3 months. Chill before enjoying and serve with garnish if desired. For a sweeter drink, add a splash of simple syrup before serving.

strawberry lemonade
HARD ICED TEA

AVERAGE ABV: 4.5 TO 5.5% ● **ABOUT 1 GALLON**

As much as I love an Arnold Palmer (page 93), when I see the first plump, ruby-red strawberries appear in the late spring, I can't help but give that classic drink a fruity upgrade. This hard iced tea is a perfect blend of sweet strawberry, puckery lemon, and mellow iced tea. It's a crowd-pleaser, for sure, so it's a good choice if you're on beverage duty for any upcoming gatherings.

 TRY THIS! Swap out the strawberries for any other fruity lemonade combo. Peach lemonade, raspberry lemonade, blueberry lemonade—they're all great!

1 gallon spring, artesian, or reverse osmosis water

12 ounces (2⅓ cups) corn sugar for fermentation, plus 1 ounce (3 tablespoons) for bottling

½ teaspoon champagne yeast

1 teaspoon yeast nutrient

10 bags good-quality black tea

½ pound (8 to 10 medium) fresh or frozen strawberries (roughly chopped if fresh)

Strips of zest and juice from 2 lemons

Lemon wedges, for garnish (optional)

Simple syrup (optional)

---

day 1

Sanitize your fermenter, air lock, and whisk. In the fermenter, combine the water, 12 ounces of the corn sugar, and the champagne yeast. Whisk until the corn sugar is dissolved and the liquid is foamy on top, 30 to 60 seconds. Seal the fermenter, fill the air lock with sanitizer, and insert it into the fermenter.

Place the fermenter somewhere dark, slightly warm (70° to 80°F), and out of the way. You should start to see signs of fermentation (like bubbling in the air lock) within 24 to 48 hours.

---

day 2

In a small, sanitized measuring cup, dissolve ½ teaspoon of the yeast nutrient in 2 tablespoons of warm water. Add to the fermenter, reseal, and replace the air lock. Swirl gently to distribute and return to your fermentation spot.

---

day 3

In a small, sanitized measuring cup, dissolve ¼ teaspoon of the yeast nutrient in 2 tablespoons of warm water. Add to the fermenter, reseal, and replace the air lock. Swirl gently to distribute and return to your fermentation spot.

CONTINUED

day 4	In a small, sanitized measuring cup, dissolve the remaining ¼ teaspoon yeast nutrient in 2 tablespoons of warm water. Add to the fermenter, reseal, and replace the air lock. Swirl gently to distribute and return to your fermentation spot.
days 5 to 14	Active fermentation will peak around Day 5 and then mostly finish around Day 7. Anytime between Day 7 and Day 10, add the tea, strawberries, and lemon zest and juice to the fermenter and infuse for at least 3 days or up to 7 days. After adding flavorings, you may see renewed signs of fermentation, though less vigorous than originally. Once you see no more signs of fermentation (like bubbles in the air lock), you can assume fermentation is complete. Wait another 24 hours to be safe, then proceed with bottling.
bottling day	When you're ready to bottle, sanitize a liquid measuring cup, spoon, large pot (1 gallon or larger), siphon, bottle filler, bottles, and caps. In the measuring cup, combine the remaining 1 ounce corn sugar with ½ cup of water and stir to dissolve. Pour this sugar water into the pot. Siphon the hard iced tea into the pot with the sugar water, leaving behind the tea bags and any solids. Attach the bottle filler to your siphon, transfer the hard iced tea into the bottles, and cap.
	Store somewhere cool, dark, and out of the way for 1 to 2 weeks to carbonate, or for up to 3 months. Chill before enjoying and serve with garnish if desired. For a sweeter drink, add a splash of simple syrup before serving.

mint julep HARD ICED TEA

AVERAGE ABV: 4.5 TO 5.5% ● **ABOUT 1 GALLON**

Swap out the fresh mint traditionally used to make mint juleps and try mint tea instead! This isn't just a gimmick—I find that mint tea provides a smoother, softer mint flavor to this bottled drink. The key is to use a really good-quality Moroccan mint tea, so don't skimp here. Pick up a nice bourbon while you're at it. You don't need to blow your whole budget, but it's worth splurging on a bottle of something good.

 TRY THIS! This might sound odd, but skip the bourbon and add 3 tablespoons of cacao nibs. They will add a chocolate undertone to this minty drink, like Thin Mints or Andes mints.

1 gallon spring, artesian, or reverse osmosis water

12 ounces (2⅓ cups) corn sugar for fermentation, plus 1 ounce (3 tablespoons) for bottling

½ teaspoon champagne yeast

½ ounce oak cubes

4 ounces (½ cup) bourbon

1 teaspoon yeast nutrient

5 bags good-quality Moroccan mint tea

5 bags good-quality black tea

Fresh mint, for garnish (optional)

Simple syrup (optional)

day 1

Sanitize your fermenter, air lock, and whisk. In the fermenter, combine the water, 12 ounces of the corn sugar, and the champagne yeast. Whisk until the corn sugar is dissolved and the liquid is foamy on top, 30 to 60 seconds. Seal the fermenter, fill the air lock with sanitizer, and insert it into the fermenter.

Place the fermenter somewhere dark, slightly warm (70° to 80°F), and out of the way. You should start to see signs of fermentation (like bubbling in the air lock) within 24 to 48 hours.

Meanwhile, combine the oak cubes and bourbon in a small airtight container. Set aside for at least 5 days, shaking occasionally.

day 2

In a small, sanitized measuring cup, dissolve ½ teaspoon of the yeast nutrient in 2 tablespoons of warm water. Add to the fermenter, reseal, and replace the air lock. Swirl gently to distribute and return to your fermentation spot.

day 3

In a small, sanitized measuring cup, dissolve ¼ teaspoon of the yeast nutrient in 2 tablespoons of warm water. Add to the fermenter, reseal, and replace the air lock. Swirl gently to distribute and return to your fermentation spot.

day 4	In a small, sanitized measuring cup, dissolve the remaining ¼ teaspoon yeast nutrient in 2 tablespoons of warm water. Add to the fermenter, reseal, and replace the air lock. Swirl gently to distribute and return to your fermentation spot.
days 5 to 14	Active fermentation will peak around Day 5 and then mostly finish around Day 7. Anytime between Day 7 and Day 10, add the tea along with the oak cubes and bourbon to the fermenter and infuse for at least 3 days or up to 7 days. After adding flavorings, you may see renewed signs of fermentation, though less vigorous than originally. Once you see no more signs of fermentation (like bubbles in the air lock), you can assume fermentation is complete. Wait another 24 hours to be safe, then proceed with bottling.
bottling day	When you're ready to bottle, sanitize a liquid measuring cup, spoon, large pot (1 gallon or larger), siphon, bottle filler, bottles, and caps. In the measuring cup, combine the remaining 1 ounce corn sugar with ½ cup of water and stir to dissolve. Pour this sugar water into the pot. Siphon the hard iced tea into the pot with the sugar water, leaving behind the tea bags and any solids. Attach the bottle filler to your siphon, transfer the hard iced tea into the bottles, and cap.
	Store somewhere cool, dark, and out of the way for 1 to 2 weeks to carbonate, or for up to 3 months. Chill before enjoying and serve with garnish if desired. For a sweeter drink, add a splash of simple syrup before serving.

sangria HARD ICED TEA

AVERAGE ABV: 4.5 TO 5.5% ● **ABOUT 1 GALLON**

Instead of using the traditional sangria base of red wine, we're using a blend of pomegranate juice, cherry juice, and black tea to mimic the berry flavors and puckery tannins found in wine. You can carbonate this one or leave it still (just skip the sugar when bottling). Serve it with a splash of simple syrup and a few orange rounds tucked into the glass.

 TRY THIS! For a more robust spin, replace the water with ½ gallon of apple juice and increase the amount of pomegranate and cherry juices to a quart each.

3 quarts spring, artesian, or reverse osmosis water

16 ounces (2 cups) pomegranate juice

16 ounces (2 cups) sweet cherry juice

12 ounces (2⅓ cups) corn sugar for fermentation, plus 1 ounce (3 tablespoons) for bottling

½ teaspoon champagne yeast

1 teaspoon yeast nutrient

10 bags good-quality black tea

Strips of zest and juice from 2 oranges

14 ounces (2 medium) sweet apples, diced

Orange slices, for garnish (optional)

Simple syrup (optional)

day 1	Sanitize your fermenter, air lock, and whisk. In the fermenter, combine the water, pomegranate juice, cherry juice, 12 ounces of the corn sugar, and the champagne yeast. Whisk until the corn sugar is dissolved and the liquid is foamy on top, 30 to 60 seconds. Seal the fermenter, fill the air lock with sanitizer, and insert it into the fermenter.
	Place the fermenter somewhere dark, slightly warm (70° to 80°F), and out of the way. You should start to see signs of fermentation (like bubbling in the air lock) within 24 to 48 hours.
day 2	In a small, sanitized measuring cup, dissolve ½ teaspoon of the yeast nutrient in 2 tablespoons of warm water. Add to the fermenter, reseal, and replace the air lock. Swirl gently to distribute and return to your fermentation spot.
day 3	In a small, sanitized measuring cup, dissolve ¼ teaspoon of the yeast nutrient in 2 tablespoons of warm water. Add to the fermenter, reseal, and replace the air lock. Swirl gently to distribute and return to your fermentation spot.

CONTINUED

day 4	In a small, sanitized measuring cup, dissolve the remaining ¼ teaspoon yeast nutrient in 2 tablespoons of warm water. Add to the fermenter, reseal, and replace the air lock. Swirl gently to distribute and return to your fermentation spot.
days 5 to 14	Active fermentation will peak around Day 5 and then mostly finish around Day 7. Anytime between Day 7 and Day 10, add the tea, orange zest and juice, and apples to the fermenter and infuse for at least 3 days or up to 7 days.
	After adding flavorings, you may see renewed signs of fermentation, though less vigorous than originally. Once you see no more signs of fermentation (like bubbles in the air lock), you can assume fermentation is complete. Wait another 24 hours to be safe, then proceed with bottling.
bottling day	When you're ready to bottle, sanitize a liquid measuring cup, spoon, large pot (1 gallon or larger), siphon, bottle filler, bottles, and caps. In the measuring cup, combine the remaining 1 ounce corn sugar with ½ cup of water and stir to dissolve. Pour this sugar water into the pot.
	Siphon the hard iced tea into the pot with the sugar water, leaving behind the tea bags and any solids. Attach the bottle filler to your siphon, transfer the hard iced tea into the bottles, and cap.

Store somewhere cool, dark, and out of the way for 1 to 2 weeks to carbonate, or for up to 3 months. Chill before enjoying and serve with garnish if desired. For a sweeter drink, add a splash of simple syrup before serving.

mai tai HARD ICED TEA

AVERAGE ABV: 5 TO 6% ● **ABOUT 1 GALLON**

Mai tai cocktails are a blend of sweet rum, orange curaçao, orgeat (an almond-based syrup), and fresh citrus. They are intensely flavored, intensely boozy, and intensely delicious. Consider this hard iced tea a gracious nod to the original tiki drink. It mirrors the classic flavors, but you can enjoy more than one in an evening without ending up dancing on the tables.

 TRY THIS! Pour your mai tai iced tea into a glass and add a ½ ounce floater of dark rum on top just before serving. This more closely mimics the original drink.

1 gallon spring, artesian, or reverse osmosis water

12 ounces (2⅓ cups) corn sugar for fermentation, plus 1 ounce (3 tablespoons) for bottling

½ teaspoon champagne yeast

½ ounce oak cubes

4 ounces (½ cup) golden rum

1 teaspoon yeast nutrient

5 bags good-quality rooibos tea

4 ounces (½ cup) curaçao

Strips of zest from 2 limes

Juice from 4 limes

Strips of zest and juice from 2 oranges

1 tablespoon almond extract

Orange slices, pineapple wedges, and maraschino cherries, for garnish (optional)

Simple syrup (optional)

---

day 1

Sanitize your fermenter, air lock, and whisk. In the fermenter, combine the water, 12 ounces of the corn sugar, and the champagne yeast. Whisk until the corn sugar is dissolved and the liquid is foamy on top, 30 to 60 seconds. Seal the fermenter, fill the air lock with sanitizer, and insert it into the fermenter.

Place the fermenter somewhere dark, slightly warm (70° to 80°F), and out of the way. You should start to see signs of fermentation (like bubbling in the air lock) within 24 to 48 hours.

Meanwhile, combine the oak cubes and rum in a small airtight container. Set aside for at least 5 days, shaking occasionally.

---

day 2

In a small, sanitized measuring cup, dissolve ½ teaspoon of the yeast nutrient in 2 tablespoons of warm water. Add to the fermenter, reseal, and replace the air lock. Swirl gently to distribute and return to your fermentation spot.

CONTINUED

day 3	In a small, sanitized measuring cup, dissolve ¼ teaspoon of the yeast nutrient in 2 tablespoons of warm water. Add to the fermenter, reseal, and replace the air lock. Swirl gently to distribute and return to your fermentation spot.
day 4	In a small, sanitized measuring cup, dissolve the remaining ¼ teaspoon yeast nutrient in 2 tablespoons of warm water. Add to the fermenter, reseal, and replace the air lock. Swirl gently to distribute and return to your fermentation spot.
days 5 to 14	Active fermentation will peak around Day 5 and then mostly finish around Day 7. Anytime between Day 7 and Day 10, add the tea along with the oak cubes and rum, curaçao, citrus zest and juice, and almond extract to the fermenter and infuse for at least 3 days or up to 7 days. After adding flavorings, you may see renewed signs of fermentation, though less vigorous than originally. Once you see no more signs of fermentation (like bubbles in the air lock), you can assume fermentation is complete. Wait another 24 hours to be safe, then proceed with bottling.
bottling day	When you're ready to bottle, sanitize a liquid measuring cup, spoon, large pot (1 gallon or larger), siphon, bottle filler, bottles, and caps. In the measuring cup, combine the remaining 1 ounce corn sugar with ½ cup of water and stir to dissolve. Pour this sugar water into the pot. Siphon the hard iced tea into the pot with the sugar water, leaving behind the tea bags and any solids. Using a sanitized wine thief or measuring cup, scoop out a little liquid and give it a taste. Add more rum, curaçao, citrus juice, or almond extract until the desired flavor is reached. Attach the bottle filler to your siphon, transfer the hard iced tea into the bottles, and cap.
	Store somewhere cool, dark, and out of the way for 1 to 2 weeks to carbonate, or for up to 3 months. Chill before enjoying and serve with garnish if desired. For a sweeter drink, add a splash of simple syrup before serving.

3

HARD
Kombuchas

Kombucha is inherently tricksy, and hard kombucha even more so. It's not that it's technically all that difficult—the actual process isn't very different from the other drinks in this book—but the nature of kombucha makes it . . . well, tricksy.

Kombucha is made by fermenting sweetened tea with a scoby, an acronym for "symbiotic culture of bacteria and yeast." But the yeast and bacteria aren't the domesticated, predictable microorganisms that are grown in labs and sold in neatly sealed packets. They are unruly, capricious, and literally wild—they are born from the bacteria and yeast that hang out around your home and are responsible for giving kombucha its zippy flavor and flamboyant fizz. This wildness makes it simultaneously very exciting to brew kombucha and occasionally very annoying when they don't behave as expected.

To ensure maximum excitement and minimum annoyance, start by making sure that your scoby is healthy and happy: If you're not already a regular kombucha brewer, make at least four batches of regular, nonalcoholic Everyday Kombucha (page 108) over the course of about a month (7 to 10 days between each batch). This gets your scoby into a regular schedule, and it also prevents the kombucha from becoming too tart.

Once this is going smoothly, make your hard kombucha, which is as simple as using a batch of Everyday Kombucha as your base, adding some extra sugar and commercial yeast, and letting this go through another round of fermentation.

Why add extra yeast if we already have wild yeast in the scoby? A few reasons: to add some predictability back into the equation, because wild yeast tends to ferment more slowly than we'd like, and to budge out the bacteria from the scoby, which tends to make hard kombucha overly vinegary if left unchecked. To give ourselves even more of a leg up, I like to make a yeast starter—a mini-batch of sugar, water, and yeast—to make sure the new yeast is strong and ready to work once thrown into the ring with the wild yeast and bacteria.

What to Expect: Expect some unpredictability. When everything goes according to plan, I've found that homemade hard kombucha is essentially identical to what you buy in stores. But even if you follow the recipe to the letter and watch your kombucha with an eagle eye, some batches still turn out a little funkier than others. But that's all part of being a homebrewer. It's an adventure!

everyday KOMBUCHA

AVERAGE ABV: >2% ● **ABOUT 1 GALLON**

This recipe for basic, everyday kombucha can be enjoyed on its own, but it's also the base for all the hard kombucha recipes in this chapter! A strong, healthy scoby is essential for making delicious kombucha (boozy or otherwise), so if you're new to making kombucha or haven't been making it regularly, prepare at least four batches of this basic nonalcoholic kombucha in a row with 7 to 10 days between each batch before trying your hand at the hard version. Once your scoby is reliably growing a new thin layer with every batch and the flavor of the brew is a pleasant balance of sweet and slightly tart, you're ready to move on to hard kombucha.

TRY THIS! Any type of tea can be used to make kombucha as long as it is caffeinated and does not contain any essential oils or flavoring extracts. You can play with this recipe by going with all black tea or all green tea, or swap in white tea or pu-erh!

7 ounces (1 cup) granulated sugar

4 bags black tea, or 2 tablespoons loose-leaf black tea

4 bags green tea, or 2 tablespoons loose-leaf green tea

16 ounces (2 cups) prepared unflavored kombucha, homemade or store-bought

1 scoby

---

day 1

In a medium pot, bring 1 quart of water to a simmer. Add the sugar, black tea, and green tea, and stir until the sugar is dissolved. Remove from the heat and steep for 15 minutes. Stir in another 2½ quarts of water (to make 3½ quarts total) and cool to room temperature. Remove the tea bags and stir in the prepared unflavored kombucha.

Transfer the sweet tea mixture to a clean 1-gallon wide-mouthed glass jar. Add the scoby (it may sink, bob in the middle, or float on top—all are fine). Cover the jar with a clean, tightly woven cloth (like a napkin) and secure it with a rubber band. Place the kombucha somewhere dark, slightly warm (70° to 80°F), and out of the way.

day 2 and after

If you're making hard kombucha, ferment for at least 7 days or up to 10 days; it should still taste fairly sweet at this point. If you're making kombucha for everyday drinking, ferment at least 7 days or up to 1 month, until it tastes good to you.

When the kombucha is ready, remove the scoby (including any new layers) and 2 cups of the kombucha, and set these aside in a cloth-covered glass jar; use right away to make your next batch or store at room temperature for 1 to 4 weeks (top off with sweet tea if the liquid evaporates). The remaining kombucha can be used to make hard kombucha (page 110) or consumed as is.

If you'd like to flavor your everyday kombucha, add the flavorings (any from this book!), cover with a cloth secured with a rubber band, and let stand for 1 to 2 days. Strain and bottle in clean swing-top bottles or empty plastic soda bottles. Let stand for another 2 to 3 days to carbonate, then refrigerate before drinking. Drink within 1 month.

basic HARD KOMBUCHA

AVERAGE ABV: 7 TO 8% ● **ABOUT 1 GALLON**

Once you're ready to turn your Everyday Kombucha into hard kombucha, there's one more thing you need: a yeast starter. Mix this mini-batch of sugar, water, and champagne yeast and let it sit a few hours to make sure the champagne yeast is fully active and ready to go before combining it with the Everyday Kombucha. This ensures a strong fermentation over the next few weeks and prevents the wild yeast and bacteria from getting too squirrelly. The result will be a semisweet, slightly tangy hard kombucha that's great on its own or can be mixed with any flavorings your heart desires.

 TRY THIS! Instead of using water in your yeast starter, use any fruit juice that you like. Pick one that pairs well with any other fruits or flavorings that you plan on adding.

12 ounces (2⅓ cups) corn sugar for fermentation, plus 1 ounce (3 tablespoons) for bottling

1 teaspoon champagne yeast

3½ quarts Everyday Kombucha (page 108)

1 teaspoon yeast nutrient

Flavorings to taste

Simple syrup (optional)

---

day 1

Two to 8 hours before you plan to make your hard kombucha, prepare a yeast starter to ensure vigorous fermentation: Sanitize a 1-quart (or larger) jar and a spoon. In the jar, combine 2 cups of water, 2 ounces of the corn sugar, and the champagne yeast. Stir with the spoon until the sugar dissolves.

Cover with plastic wrap secured with a rubber band and let stand for 2 to 8 hours. You should see the plastic puff up and a layer of foam build on the surface of the liquid; both are signs of good fermentation.

When you're ready to make the hard kombucha, sanitize your fermenter, air lock, and whisk. In the fermenter, combine 10 ounces of the corn sugar, the everyday kombucha, and the yeast starter that you made. Whisk gently until the corn sugar is dissolved and the liquid is foamy on top, 30 to 60 seconds. Seal the fermenter, fill the air lock with sanitizer, and insert it into the fermenter.

Place the fermenter somewhere dark, slightly warm (70° to 80°F), and out of the way. You should start to see signs of fermentation (like bubbling in the air lock) within 24 to 48 hours.

---

day 2	In a small, sanitized measuring cup, dissolve ½ teaspoon of the yeast nutrient in 2 tablespoons of warm water. Add to the fermenter, reseal, and replace the air lock. Swirl gently to distribute and return to your fermentation spot.
day 3	In a small, sanitized measuring cup, dissolve ¼ teaspoon of the yeast nutrient in 2 tablespoons of warm water. Add to the fermenter, reseal, and replace the air lock. Swirl gently to distribute and return to your fermentation spot.
day 4	In a small, sanitized measuring cup, dissolve the remaining ¼ teaspoon yeast nutrient in 2 tablespoons of warm water. Add to the fermenter, reseal, and replace the air lock. Swirl gently to distribute and return to your fermentation spot.
days 5 to 14	Active fermentation will peak around Day 5 and then mostly finish around Day 7. Anytime between Day 7 and Day 10, add any flavorings to the fermenter and infuse for at least 3 days or up to 7 days. After adding flavorings, you may see renewed signs of fermentation, though less vigorous than originally. Once you see no more signs of fermentation, you can assume fermentation is complete (with kombucha, a few static bubbles floating in the air lock are fine as long as it's not actively bubbling). Wait another 24 hours to be safe, then proceed with bottling.
bottling day	When you're ready to bottle, sanitize a liquid measuring cup, spoon, large pot (1 gallon or larger), siphon, bottle filler, bottles, and caps. In the measuring cup, combine the remaining 1 ounce corn sugar with ½ cup of water and stir to dissolve. Pour this sugar water into the pot. Siphon the kombucha into the pot with the sugar water, leaving behind any solids. Attach the bottle filler to your siphon, transfer the hard kombucha into the bottles, and cap.
	Store somewhere cool, dark, and out of the way for 1 to 2 weeks to carbonate, and then refrigerate all the bottles. Since hard kombucha contains wild yeast, fermentation can sometimes continue after bottling and cause bottles to eventually burst if not refrigerated. Open bottles carefully outdoors or over a sink in case of gushing. Chill before enjoying. For a sweeter drink, add a splash of simple syrup before serving. Drink within 3 months.

ginger-lemon HARD KOMBUCHA

AVERAGE ABV: 7 TO 8% ○ **ABOUT 1 GALLON**

Ginger goes with kombucha the way peanut butter goes with jelly—it just plain works. In fact, I was hard-pressed not to add ginger to every recipe in this chapter. It brings a sweet heat that you can feel tingling all the way to your toes. Throw in some fresh lemon and you have a zesty, zingy brew that is impossible to resist.

TRY THIS! Add a cup of dried hibiscus flowers for a fruity flavor and a lovely rose color.

12 ounces (2⅓ cups) corn sugar for fermentation, plus 1 ounce (3 tablespoons) for bottling

1 teaspoon champagne yeast

3½ quarts Everyday Kombucha (page 108)

1 teaspoon yeast nutrient

Strips of zest and juice from 2 lemons

1 ounce (¼ cup) roughly chopped ginger

Lemon wedges, for garnish (optional)

Simple syrup (optional)

---

day 1

Two to 8 hours before you plan to make your hard kombucha, prepare a yeast starter to ensure vigorous fermentation: Sanitize a 1-quart (or larger) jar and a spoon. In the jar, combine 2 cups of water, 2 ounces of the corn sugar, and the champagne yeast. Stir with the spoon until the sugar dissolves.

Cover with plastic wrap secured with a rubber band and let stand for 2 to 8 hours. You should see the plastic puff up and a layer of foam build on the surface of the liquid; both are signs of good fermentation.

When you're ready to make the hard kombucha, sanitize your fermenter, air lock, and whisk. In the fermenter, combine 10 ounces of the corn sugar, the everyday kombucha, and the yeast starter that you made. Whisk gently until the corn sugar is dissolved and the liquid is foamy on top, 30 to 60 seconds. Seal the fermenter, fill the air lock with sanitizer, and insert it into the fermenter.

Place the fermenter somewhere dark, slightly warm (70° to 80°F), and out of the way. You should start to see signs of fermentation (like bubbling in the air lock) within 24 to 48 hours.

---

day 2

In a small, sanitized measuring cup, dissolve ½ teaspoon of the yeast nutrient in 2 tablespoons of warm water. Add to the fermenter, reseal, and replace the air lock. Swirl gently to distribute and return to your fermentation spot.

---

day 3	In a small, sanitized measuring cup, dissolve ¼ teaspoon of the yeast nutrient in 2 tablespoons of warm water. Add to the fermenter, reseal, and replace the air lock. Swirl gently to distribute and return to your fermentation spot.
day 4	In a small, sanitized measuring cup, dissolve the remaining ¼ teaspoon yeast nutrient in 2 tablespoons of warm water. Add to the fermenter, reseal, and replace the air lock. Swirl gently to distribute and return to your fermentation spot.
days 5 to 14	Active fermentation will peak around Day 5 and then mostly finish around Day 7. Anytime between Day 7 and Day 10, add the lemon zest and juice along with the ginger to the fermenter and infuse for at least 3 days or up to 7 days. After adding flavorings, you may see renewed signs of fermentation, though less vigorous than originally. Once you see no more signs of fermentation, you can assume fermentation is complete (with kombucha, a few static bubbles floating in the air lock are fine as long as it's not actively bubbling). Wait another 24 hours to be safe, then proceed with bottling.
bottling day	When you're ready to bottle, sanitize a liquid measuring cup, spoon, large pot (1 gallon or larger), siphon, bottle filler, bottles, and caps. In the measuring cup, combine the remaining 1 ounce corn sugar with ½ cup of water and stir to dissolve. Pour this sugar water into the pot. Siphon the kombucha into the pot with the sugar water, leaving behind any solids. Attach the bottle filler to your siphon, transfer the hard kombucha into the bottles, and cap.
	Store somewhere cool, dark, and out of the way for 1 to 2 weeks to carbonate, and then refrigerate all the bottles. Since hard kombucha contains wild yeast, fermentation can sometimes continue after bottling and cause bottles to eventually burst if not refrigerated. Open bottles carefully outdoors or over a sink in case of gushing. Chill before enjoying and serve with garnish if desired. For a sweeter drink, add a splash of simple syrup before serving. Drink within 3 months.

blueberry-pomegranate
HARD KOMBUCHA

AVERAGE ABV: 7 TO 8% ● **ABOUT 1 GALLON**

On its own, I find that pomegranate juice can be a tad aggressive. But partnering it with blueberries tempers its tart, tannic qualities and brings some fruity sweetness to the party. Every sip is a satisfying mix of each fruit's best qualities. Note that this recipe uses the pomegranate juice to make the yeast starter, so no extra water is needed.

▬ **TRY THIS!** For an extra kick, add ¼ cup roughly chopped ginger along with the blueberries.

16 ounces (2 cups) pomegranate juice

12 ounces (2⅓ cups) corn sugar for fermentation, plus 1 ounce (3 tablespoons) for bottling

1 teaspoon champagne yeast

3½ quarts Everyday Kombucha (page 108)

1 teaspoon yeast nutrient

1 pound (about 3½ cups) fresh or frozen blueberries (lightly muddled if fresh)

Blueberries, for garnish (optional)

Simple syrup (optional)

---

day 1

Two to 8 hours before you plan to make your hard kombucha, prepare a yeast starter to ensure vigorous fermentation: Sanitize a 1-quart (or larger) jar and a spoon. In the jar, combine the pomegranate juice, 2 ounces of the corn sugar, and the champagne yeast. Stir with the spoon until the sugar dissolves.

Cover with plastic wrap secured with a rubber band and let stand for 2 to 8 hours. You should see the plastic puff up and a layer of foam build on the surface of the liquid; both are signs of good fermentation.

When you're ready to make the hard kombucha, sanitize your fermenter, air lock, and whisk. In the fermenter, combine 10 ounces of the corn sugar, the everyday kombucha, and the yeast starter that you made. Whisk gently until the corn sugar is dissolved and the liquid is foamy on top, 30 to 60 seconds. Seal the fermenter, fill the air lock with sanitizer, and insert it into the fermenter.

Place the fermenter somewhere dark, slightly warm (70° to 80°F), and out of the way. You should start to see signs of fermentation (like bubbling in the air lock) within 24 to 48 hours.

CONTINUED

day 2	In a small, sanitized measuring cup, dissolve ½ teaspoon of the yeast nutrient in 2 tablespoons of warm water. Add to the fermenter, reseal, and replace the air lock. Swirl gently to distribute and return to your fermentation spot.
day 3	In a small, sanitized measuring cup, dissolve ¼ teaspoon of the yeast nutrient in 2 tablespoons of warm water. Add to the fermenter, reseal, and replace the air lock. Swirl gently to distribute and return to your fermentation spot.
day 4	In a small, sanitized measuring cup, dissolve the remaining ¼ teaspoon yeast nutrient in 2 tablespoons of warm water. Add to the fermenter, reseal, and replace the air lock. Swirl gently to distribute and return to your fermentation spot.
days 5 to 14	Active fermentation will peak around Day 5 and then mostly finish around Day 7. Anytime between Day 7 and Day 10, add the blueberries to the fermenter and infuse for at least 3 days or up to 7 days. After adding flavorings, you may see renewed signs of fermentation, though less vigorous than originally. Once you see no more signs of fermentation, you can assume fermentation is complete (with kombucha, a few static bubbles floating in the air lock are fine as long as it's not actively bubbling). Wait another 24 hours to be safe, then proceed with bottling.
bottling day	When you're ready to bottle, sanitize a liquid measuring cup, spoon, large pot (1 gallon or larger), siphon, bottle filler, bottles, and caps. In the measuring cup, combine the remaining 1 ounce corn sugar with ½ cup of water and stir to dissolve. Pour this sugar water into the pot. Siphon the kombucha into the pot with the sugar water, leaving behind any solids. Attach the bottle filler to your siphon, transfer the hard kombucha into the bottles, and cap.
	Store somewhere cool, dark, and out of the way for 1 to 2 weeks to carbonate, and then refrigerate all the bottles. Since hard kombucha contains wild yeast, fermentation can sometimes continue after bottling and cause bottles to eventually burst if not refrigerated. Open bottles carefully outdoors or over a sink in case of gushing. Chill before enjoying and serve with garnish if desired. For a sweeter drink, add a splash of simple syrup before serving. Drink within 3 months.

tepache HARD KOMBUCHA

AVERAGE ABV: 7 TO 8% ● **ABOUT 1 GALLON**

Tepache is a fizzy Mexican drink made by mixing pineapple rinds and cores with water and letting it ferment naturally with the wild yeasts and bacteria present on the pineapple. In this way, it shares a lot of similarities with kombucha, and combining them into one beverage feels like doubling up on a good thing. For maximum flavor and to avoid introducing outside yeast and bacteria, use fresh or frozen pineapple instead of the traditional rinds and cores.

 TRY THIS! I love adding other tropical flavors to this tepache base. Along with the pineapple, try adding some mango, passion fruit, guava, or all three!

12 ounces (2⅓ cups) corn sugar for fermentation, plus 1 ounce (3 tablespoons) for bottling

1 teaspoon champagne yeast

3½ quarts Everyday Kombucha (page 108)

1 teaspoon yeast nutrient

1 pound (about 3½ cups) fresh or frozen pineapple chunks

2 whole cinnamon sticks

5 whole cloves

Fresh or dried pineapple wedges, for garnish (optional)

Simple syrup (optional)

---

day 1

Two to 8 hours before you plan to make your hard kombucha, prepare a yeast starter to ensure vigorous fermentation: Sanitize a 1-quart (or larger) jar and a spoon. In the jar, combine 2 cups of water, 2 ounces of the corn sugar, and the champagne yeast. Stir with the spoon until the sugar dissolves.

Cover with plastic wrap secured with a rubber band and let stand for 2 to 8 hours. You should see the plastic puff up and a layer of foam build on the surface of the liquid; both are signs of good fermentation.

When you're ready to make the hard kombucha, sanitize your fermenter, air lock, and whisk. In the fermenter, combine 10 ounces of the corn sugar, the everyday kombucha, and the yeast starter that you made. Whisk gently until the corn sugar is dissolved and the liquid is foamy on top, 30 to 60 seconds. Seal the fermenter, fill the air lock with sanitizer, and insert it into the fermenter.

Place the fermenter somewhere dark, slightly warm (70° to 80°F), and out of the way. You should start to see signs of fermentation (like bubbling in the air lock) within 24 to 48 hours.

CONTINUED

day 2	In a small, sanitized measuring cup, dissolve ½ teaspoon of the yeast nutrient in 2 tablespoons of warm water. Add to the fermenter, reseal, and replace the air lock. Swirl gently to distribute and return to your fermentation spot.
day 3	In a small, sanitized measuring cup, dissolve ¼ teaspoon of the yeast nutrient in 2 tablespoons of warm water. Add to the fermenter, reseal, and replace the air lock. Swirl gently to distribute and return to your fermentation spot.
day 4	In a small, sanitized measuring cup, dissolve the remaining ¼ teaspoon yeast nutrient in 2 tablespoons of warm water. Add to the fermenter, reseal, and replace the air lock. Swirl gently to distribute and return to your fermentation spot.
days 5 to 14	Active fermentation will peak around Day 5 and then mostly finish around Day 7. Anytime between Day 7 and Day 10, add the pineapple, cinnamon, and cloves to the fermenter and infuse for at least 3 days or up to 7 days. After adding flavorings, you may see renewed signs of fermentation, though less vigorous than originally. Once you see no more signs of fermentation, you can assume fermentation is complete (with kombucha, a few static bubbles floating in the air lock are fine as long as it's not actively bubbling). Wait another 24 hours to be safe, then proceed with bottling.
bottling day	When you're ready to bottle, sanitize a liquid measuring cup, spoon, large pot (1 gallon or larger), siphon, bottle filler, bottles, and caps. In the measuring cup, combine the remaining 1 ounce corn sugar with ½ cup of water and stir to dissolve. Pour this sugar water into the pot. Siphon the kombucha into the pot with the sugar water, leaving behind any solids. Attach the bottle filler to your siphon, transfer the hard kombucha into the bottles, and cap.
	Store somewhere cool, dark, and out of the way for 1 to 2 weeks to carbonate, and then refrigerate all the bottles. Since hard kombucha contains wild yeast, fermentation can sometimes continue after bottling and cause bottles to eventually burst if not refrigerated. Open bottles carefully outdoors or over a sink in case of gushing. Chill before enjoying and serve with garnish if desired. For a sweeter drink, add a splash of simple syrup before serving. Drink within 3 months.

berry cherry HARD KOMBUCHA

AVERAGE ABV: 7 TO 8% ● **ABOUT 1 GALLON**

This hard kombucha is an ode to the fruit salad I grew up eating every night. It was one of the few dishes my brother and I would reliably eat, so it was always on the table. The specific fruits in the salad would change with the seasons, but my absolute favorite was whenever my mom added sweet cherries. I loved— and still love—these plump, jewel-toned little orbs, particularly when paired with early summer fruits like blueberries and raspberries.

 TRY THIS! Mix and match your favorite fruits to create your own fruit salad kombucha! Just stick to about 1¼ pounds total fruit.

12 ounces (2⅓ cups) corn sugar for fermentation, plus 1 ounce (3 tablespoons) for bottling

1 teaspoon champagne yeast

3½ quarts Everyday Kombucha (page 108)

1 teaspoon yeast nutrient

½ pound (about 1½ cups) pitted fresh or frozen cherries (sweet, or mix of sweet and tart), roughly chopped

½ pound (about 2 scant cups) fresh or frozen blueberries (lightly muddled if fresh)

¼ pound (about 1 cup) fresh or frozen raspberries or blackberries (lightly muddled if fresh)

Cherries, blueberries, or raspberries, for garnish (optional)

Simple syrup (optional)

---

day 1

Two to 8 hours before you plan to make your hard kombucha, prepare a yeast starter to ensure vigorous fermentation: Sanitize a 1-quart (or larger) jar and a spoon. In the jar, combine 2 cups of water, 2 ounces of the corn sugar, and the champagne yeast. Stir with the spoon until the sugar dissolves.

Cover with plastic wrap secured with a rubber band and let stand for 2 to 8 hours. You should see the plastic puff up and a layer of foam build on the surface of the liquid; both are signs of good fermentation.

When you're ready to make the hard kombucha, sanitize your fermenter, air lock, and whisk. In the fermenter, combine 10 ounces of the corn sugar, the everyday kombucha, and the yeast starter that you made. Whisk gently until the corn sugar is dissolved and the liquid is foamy on top, 30 to 60 seconds. Seal the fermenter, fill the air lock with sanitizer, and insert it into the fermenter.

Place the fermenter somewhere dark, slightly warm (70° to 80°F), and out of the way. You should start to see signs of fermentation (like bubbling in the air lock) within 24 to 48 hours.

CONTINUED

day 2	In a small, sanitized measuring cup, dissolve ½ teaspoon of the yeast nutrient in 2 tablespoons of warm water. Add to the fermenter, reseal, and replace the air lock. Swirl gently to distribute and return to your fermentation spot.
day 3	In a small, sanitized measuring cup, dissolve ¼ teaspoon of the yeast nutrient in 2 tablespoons of warm water. Add to the fermenter, reseal, and replace the air lock. Swirl gently to distribute and return to your fermentation spot.
day 4	In a small, sanitized measuring cup, dissolve the remaining ¼ teaspoon yeast nutrient in 2 tablespoons of warm water. Add to the fermenter, reseal, and replace the air lock. Swirl gently to distribute and return to your fermentation spot.
days 5 to 14	Active fermentation will peak around Day 5 and then mostly finish around Day 7. Anytime between Day 7 and Day 10, add the cherries, blueberries, and raspberries to the fermenter and infuse for at least 3 days or up to 7 days. After adding flavorings, you may see renewed signs of fermentation, though less vigorous than originally. Once you see no more signs of fermentation, you can assume fermentation is complete (with kombucha, a few static bubbles floating in the air lock are fine as long as it's not actively bubbling). Wait another 24 hours to be safe, then proceed with bottling.
bottling day	When you're ready to bottle, sanitize a liquid measuring cup, spoon, large pot (1 gallon or larger), siphon, bottle filler, bottles, and caps. In the measuring cup, combine the remaining 1 ounce corn sugar with ½ cup of water and stir to dissolve. Pour this sugar water into the pot. Siphon the kombucha into the pot with the sugar water, leaving behind any solids. Attach the bottle filler to your siphon, transfer the hard kombucha into the bottles, and cap.
	Store somewhere cool, dark, and out of the way for 1 to 2 weeks to carbonate, and then refrigerate all the bottles. Since hard kombucha contains wild yeast, fermentation can sometimes continue after bottling and cause bottles to eventually burst if not refrigerated. Open bottles carefully outdoors or over a sink in case of gushing. Chill before enjoying and serve with garnish if desired. For a sweeter drink, add a splash of simple syrup before serving. Drink within 3 months.

strawberry HARD KOMBUCHA

AVERAGE ABV: 7 TO 8% ● **ABOUT 1 GALLON**

I could happily drink strawberry kombucha every single day, boozy or otherwise. The sweetness of juicy ripe strawberries and the tanginess of kombucha make such a refreshing, energizing duo—trust me on this one. I almost always use frozen strawberries for this recipe so I can make it all year round, but if you have a glut of fresh strawberries on your hands, I can think of no better way to put them to use.

 TRY THIS! Swap out half the strawberries for another fruit friend, like peaches, blueberries, or (another personal favorite) rhubarb.

---

12 ounces (2⅓ cups) corn sugar for fermentation, plus 1 ounce (3 tablespoons) for bottling

1 teaspoon champagne yeast

3½ quarts Everyday Kombucha (page 108)

1 teaspoon yeast nutrient

2 pounds (32 to 40 medium) fresh or frozen strawberries (roughly chopped if fresh)

Sliced strawberries, for garnish (optional)

Simple syrup (optional)

---

day 1

Two to 8 hours before you plan to make your hard kombucha, prepare a yeast starter to ensure vigorous fermentation: Sanitize a 1-quart (or larger) jar and a spoon. In the jar, combine 2 cups of water, 2 ounces of the corn sugar, and the champagne yeast. Stir with the spoon until the sugar dissolves.

Cover with plastic wrap secured with a rubber band and let stand for 2 to 8 hours. You should see the plastic puff up and a layer of foam build on the surface of the liquid; both are signs of good fermentation.

When you're ready to make the hard kombucha, sanitize your fermenter, air lock, and whisk. In the fermenter, combine 10 ounces of the corn sugar, the everyday kombucha, and the yeast starter that you made. Whisk gently until the corn sugar is dissolved and the liquid is foamy on top, 30 to 60 seconds. Seal the fermenter, fill the air lock with sanitizer, and insert it into the fermenter.

Place the fermenter somewhere dark, slightly warm (70° to 80°F), and out of the way. You should start to see signs of fermentation (like bubbling in the air lock) within 24 to 48 hours.

CONTINUED

day 2	In a small, sanitized measuring cup, dissolve ½ teaspoon of the yeast nutrient in 2 tablespoons of warm water. Add to the fermenter, reseal, and replace the air lock. Swirl gently to distribute and return to your fermentation spot.
day 3	In a small, sanitized measuring cup, dissolve ¼ teaspoon of the yeast nutrient in 2 tablespoons of warm water. Add to the fermenter, reseal, and replace the air lock. Swirl gently to distribute and return to your fermentation spot.
day 4	In a small, sanitized measuring cup, dissolve the remaining ¼ teaspoon yeast nutrient in 2 tablespoons of warm water. Add to the fermenter, reseal, and replace the air lock. Swirl gently to distribute and return to your fermentation spot.
days 5 to 14	Active fermentation will peak around Day 5 and then mostly finish around Day 7. Anytime between Day 7 and Day 10, add the strawberries to the fermenter and infuse for at least 3 days or up to 7 days. After adding flavorings, you may see renewed signs of fermentation, though less vigorous than originally. Once you see no more signs of fermentation, you can assume fermentation is complete (with kombucha, a few static bubbles floating in the air lock are fine as long as it's not actively bubbling). Wait another 24 hours to be safe, then proceed with bottling.
bottling day	When you're ready to bottle, sanitize a liquid measuring cup, spoon, large pot (1 gallon or larger), siphon, bottle filler, bottles, and caps. In the measuring cup, combine the remaining 1 ounce corn sugar with ½ cup of water and stir to dissolve. Pour this sugar water into the pot. Siphon the kombucha into the pot with the sugar water, leaving behind any solids. Attach the bottle filler to your siphon, transfer the hard kombucha into the bottles, and cap.
	Store somewhere cool, dark, and out of the way for 1 to 2 weeks to carbonate, and then refrigerate all the bottles. Since hard kombucha contains wild yeast, fermentation can sometimes continue after bottling and cause bottles to eventually burst if not refrigerated. Open bottles carefully outdoors or over a sink in case of gushing. Chill before enjoying and serve with garnish if desired. For a sweeter drink, add a splash of simple syrup before serving. Drink within 3 months.

mangonada HARD KOMBUCHA

AVERAGE ABV: 7 TO 8% ● **ABOUT 1 GALLON**

If you haven't yet had the pleasure, a mangonada is a Mexican treat made by blending mangos and lime to a slushy-like consistency and then serving with a generous swirl of chamoy (a puree of pickled fruit, chiles, and spices that you can find at Mexican groceries or online). The result is a sweet, tangy, spicy summertime staple. This booch-ified version lends itself to taco bar parties and backyard hangs around the grill.

TRY THIS! Add ¼ cup of blanco or añejo tequila for an extra kick.

12 ounces (2⅓ cups) corn sugar for fermentation, plus 1 ounce (3 tablespoons) for bottling

1 teaspoon champagne yeast

3½ quarts Everyday Kombucha (page 108)

1 teaspoon yeast nutrient

1 pound (about 3¼ cups) fresh or frozen mango chunks

2½ ounces (¼ cup) chamoy

Zest and juice from 2 limes

Tajin, to rim the glass (optional)

Mango cubes or lime wedges, for garnish (optional)

Simple syrup (optional)

Two to 8 hours before you plan to make your hard kombucha, prepare a yeast starter to ensure vigorous fermentation: Sanitize a 1-quart (or larger) jar and a spoon. In the jar, combine 2 cups of water, 2 ounces of the corn sugar, and the champagne yeast. Stir with the spoon until the sugar dissolves.

Cover with plastic wrap secured with a rubber band and let stand for 2 to 8 hours. You should see the plastic puff up and a layer of foam build on the surface of the liquid; both are signs of good fermentation.

day 1

When you're ready to make the hard kombucha, sanitize your fermenter, air lock, and whisk. In the fermenter, combine 10 ounces of the corn sugar, the everyday kombucha, and the yeast starter that you made. Whisk gently until the corn sugar is dissolved and the liquid is foamy on top, 30 to 60 seconds. Seal the fermenter, fill the air lock with sanitizer, and insert it into the fermenter.

Place the fermenter somewhere dark, slightly warm (70° to 80°F), and out of the way. You should start to see signs of fermentation (like bubbling in the air lock) within 24 to 48 hours.

CONTINUED

day 2	In a small, sanitized measuring cup, dissolve ½ teaspoon of the yeast nutrient in 2 tablespoons of warm water. Add to the fermenter, reseal, and replace the air lock. Swirl gently to distribute and return to your fermentation spot.
day 3	In a small, sanitized measuring cup, dissolve ¼ teaspoon of the yeast nutrient in 2 tablespoons of warm water. Add to the fermenter, reseal, and replace the air lock. Swirl gently to distribute and return to your fermentation spot.
day 4	In a small, sanitized measuring cup, dissolve the remaining ¼ teaspoon yeast nutrient in 2 tablespoons of warm water. Add to the fermenter, reseal, and replace the air lock. Swirl gently to distribute and return to your fermentation spot.
days 5 to 14	Active fermentation will peak around Day 5 and then mostly finish around Day 7. Anytime between Day 7 and Day 10, add the mango, chamoy, and lime zest and juice to the fermenter and infuse for at least 3 days or up to 7 days. After 3 days of infusing, use a sanitized wine thief or measuring cup to scoop out a little liquid and give it a taste. Add more chamoy or lime juice until the desired flavor and level of spiciness are reached. After adding flavorings, you may see renewed signs of fermentation, though less vigorous than originally. Once you see no more signs of fermentation, you can assume fermentation is complete (with kombucha, a few static bubbles floating in the air lock are fine as long as it's not actively bubbling). Wait another 24 hours to be safe, then proceed with bottling.
bottling day	When you're ready to bottle, sanitize a liquid measuring cup, spoon, large pot (1 gallon or larger), siphon, bottle filler, bottles, and caps. In the measuring cup, combine the remaining 1 ounce corn sugar with ½ cup of water and stir to dissolve. Pour this sugar water into the pot. Siphon the kombucha into the pot with the sugar water, leaving behind any solids. Attach the bottle filler to your siphon, transfer the hard kombucha into the bottles, and cap.
	Store somewhere cool, dark, and out of the way for 1 to 2 weeks to carbonate, and then refrigerate all the bottles. Since hard kombucha contains wild yeast, fermentation can sometimes continue after bottling and cause bottles to eventually burst if not refrigerated. Open bottles carefully outdoors or over a sink in case of gushing. Chill before enjoying and serve in a rimmed glass with garnish if desired. For a sweeter drink, add a splash of simple syrup before serving. Drink within 3 months.

ginger mojito HARD KOMBUCHA

AVERAGE ABV: 7 TO 8% ● **ABOUT 1 GALLON**

Bright lime, zesty mint, and spicy ginger join forces to make this riff on the classic mojito cocktail. You don't even really need rum to complete the picture, but I certainly won't say no if you'd like to add it! This is a good one to serve at summer parties as an option for people who want the flavor and fun of a cocktail with less of a boozy punch.

 TRY THIS! For a version of a Dark 'n' Stormy, skip the mint and use dark rum instead of light.

12 ounces (2⅓ cups) corn sugar for fermentation, plus 1 ounce (3 tablespoons) for bottling

1 teaspoon champagne yeast

3½ quarts Everyday Kombucha (page 108)

1 teaspoon yeast nutrient

Strips of zest and juice from 3 limes

1 ounce (¼ cup) roughly chopped ginger

¼ cup packed mint leaves

2 ounces (¼ cup) light rum (optional)

Mint and lime wedges, for garnish (optional)

Simple syrup (optional)

---

day 1

Two to 8 hours before you plan to make your hard kombucha, prepare a yeast starter to ensure vigorous fermentation: Sanitize a 1-quart (or larger) jar and a spoon. In the jar, combine 2 cups of water, 2 ounces of the corn sugar, and the champagne yeast. Stir with the spoon until the sugar dissolves.

Cover with plastic wrap secured with a rubber band and let stand for 2 to 8 hours. You should see the plastic puff up and a layer of foam build on the surface of the liquid; both are signs of good fermentation.

When you're ready to make the hard kombucha, sanitize your fermenter, air lock, and whisk. In the fermenter, combine 10 ounces of the corn sugar, the everyday kombucha, and the yeast starter that you made. Whisk gently until the corn sugar is dissolved and the liquid is foamy on top, 30 to 60 seconds. Seal the fermenter, fill the air lock with sanitizer, and insert it into the fermenter.

Place the fermenter somewhere dark, slightly warm (70° to 80°F), and out of the way. You should start to see signs of fermentation (like bubbling in the air lock) within 24 to 48 hours.

CONTINUED

day 2	In a small, sanitized measuring cup, dissolve ½ teaspoon of the yeast nutrient in 2 tablespoons of warm water. Add to the fermenter, reseal, and replace the air lock. Swirl gently to distribute and return to your fermentation spot.
day 3	In a small, sanitized measuring cup, dissolve ¼ teaspoon of the yeast nutrient in 2 tablespoons of warm water. Add to the fermenter, reseal, and replace the air lock. Swirl gently to distribute and return to your fermentation spot.
day 4	In a small, sanitized measuring cup, dissolve the remaining ¼ teaspoon yeast nutrient in 2 tablespoons of warm water. Add to the fermenter, reseal, and replace the air lock. Swirl gently to distribute and return to your fermentation spot.
days 5 to 14	Active fermentation will peak around Day 5 and then mostly finish around Day 7. Anytime between Day 7 and Day 10, add the lime zest and juice, ginger, mint, and rum (if using) to the fermenter and infuse for at least 3 days or up to 7 days. After adding flavorings, you may see renewed signs of fermentation, though less vigorous than originally. Once you see no more signs of fermentation, you can assume fermentation is complete (with kombucha, a few static bubbles floating in the air lock are fine as long as it's not actively bubbling). Wait another 24 hours to be safe, then proceed with bottling.
bottling day	When you're ready to bottle, sanitize a liquid measuring cup, spoon, large pot (1 gallon or larger), siphon, bottle filler, bottles, and caps. In the measuring cup, combine the remaining 1 ounce corn sugar with ½ cup of water and stir to dissolve. Pour this sugar water into the pot. Siphon the kombucha into the pot with the sugar water, leaving behind any solids. Attach the bottle filler to your siphon, transfer the hard kombucha into the bottles, and cap.
	Store somewhere cool, dark, and out of the way for 1 to 2 weeks to carbonate, and then refrigerate all the bottles. Since hard kombucha contains wild yeast, fermentation can sometimes continue after bottling and cause bottles to eventually burst if not refrigerated. Open bottles carefully outdoors or over a sink in case of gushing. Chill before enjoying and serve with garnish if desired. For a sweeter drink, add a splash of simple syrup before serving. Drink within 3 months.

fizzy mimosa HARD KOMBUCHA

AVERAGE ABV: 7 TO 8% ● **ABOUT 1 GALLON**

A brunch-ready hard kombucha? Don't mind if I do! The simple addition of orange zest and juice transforms a batch of hard kombucha into a light, easy-drinking beverage that goes splendidly with a plate of eggs and a side of pancakes. Serve this one with a slice of fresh orange.

 TRY THIS! Add a split vanilla bean or a tablespoon of vanilla extract for a nod to an Orange Julius or swap the oranges for a pound of peaches in a nod to the Bellini cocktail.

12 ounces (2⅓ cups) corn sugar for fermentation, plus 1 ounce (3 tablespoons) for bottling

1 teaspoon champagne yeast

3½ quarts Everyday Kombucha (page 108)

1 teaspoon yeast nutrient

Strips of zest and juice from 4 oranges

Orange slices, for garnish (optional)

Simple syrup (optional)

---

day 1

Two to 8 hours before you plan to make your hard kombucha, prepare a yeast starter to ensure vigorous fermentation: Sanitize a 1-quart (or larger) jar and a spoon. In the jar, combine 2 cups of water, 2 ounces of the corn sugar, and the champagne yeast. Stir with the spoon until the sugar dissolves.

Cover with plastic wrap secured with a rubber band and let stand for 2 to 8 hours. You should see the plastic puff up and a layer of foam build on the surface of the liquid; both are signs of good fermentation.

When you're ready to make the hard kombucha, sanitize your fermenter, air lock, and whisk. In the fermenter, combine 10 ounces of the corn sugar, the everyday kombucha, and the yeast starter that you made. Whisk gently until the corn sugar is dissolved and the liquid is foamy on top, 30 to 60 seconds. Seal the fermenter, fill the air lock with sanitizer, and insert it into the fermenter.

Place the fermenter somewhere dark, slightly warm (70° to 80°F), and out of the way. You should start to see signs of fermentation (like bubbling in the air lock) within 24 to 48 hours.

---

day 2

In a small, sanitized measuring cup, dissolve ½ teaspoon of the yeast nutrient in 2 tablespoons of warm water. Add to the fermenter, reseal, and replace the air lock. Swirl gently to distribute and return to your fermentation spot.

---

day 3	In a small, sanitized measuring cup, dissolve ¼ teaspoon of the yeast nutrient in 2 tablespoons of warm water. Add to the fermenter, reseal, and replace the air lock. Swirl gently to distribute and return to your fermentation spot.
day 4	In a small, sanitized measuring cup, dissolve the remaining ¼ teaspoon yeast nutrient in 2 tablespoons of warm water. Add to the fermenter, reseal, and replace the air lock. Swirl gently to distribute and return to your fermentation spot.
days 5 to 14	Active fermentation will peak around Day 5 and then mostly finish around Day 7. Anytime between Day 7 and Day 10, add the orange zest and juice to the fermenter and infuse for at least 3 days or up to 7 days. After adding flavorings, you may see renewed signs of fermentation, though less vigorous than originally. Once you see no more signs of fermentation, you can assume fermentation is complete (with kombucha, a few static bubbles floating in the air lock are fine as long as it's not actively bubbling). Wait another 24 hours to be safe, then proceed with bottling.
bottling day	When you're ready to bottle, sanitize a liquid measuring cup, spoon, large pot (1 gallon or larger), siphon, bottle filler, bottles, and caps. In the measuring cup, combine the remaining 1 ounce corn sugar with ½ cup of water and stir to dissolve. Pour this sugar water into the pot. Siphon the kombucha into the pot with the sugar water, leaving behind any solids. Attach the bottle filler to your siphon, transfer the hard kombucha into the bottles, and cap.
	Store somewhere cool, dark, and out of the way for 1 to 2 weeks to carbonate, and then refrigerate all the bottles. Since hard kombucha contains wild yeast, fermentation can sometimes continue after bottling and cause bottles to eventually burst if not refrigerated. Open bottles carefully outdoors or over a sink in case of gushing. Chill before enjoying and serve with garnish if desired. For a sweeter drink, add a splash of simple syrup before serving. Drink within 3 months.

painkiller HARD KOMBUCHA

AVERAGE ABV: 8.5 TO 9.5% ● **ABOUT 1 GALLON**

A tiki bar staple, the Painkiller cocktail is like a more citrusy version of the piña colada. It's traditionally made with Pusser's rum, a dark high-proof rum that was distributed to sailors in the British Royal Navy until 1970. Use this rum, or any other dark, aged rum, for our hard kombucha riff. Also be sure to use unsweetened coconut flakes, not shredded coconut, for best flavor.

TRY THIS! For a piña colada hard kombucha, skip the citrus and use a lighter rum.

1 quart (4 cups) pineapple juice

12 ounces (2⅓ cups) corn sugar for fermentation, plus 1 ounce (3 tablespoons) for bottling

1 teaspoon champagne yeast

3 quarts Everyday Kombucha (page 108)

½ ounce oak cubes

4 ounces (½ cup) dark rum, such as Pusser's

1 teaspoon yeast nutrient

Strips of zest and juice from 2 oranges

Strips of zest and juice from 1 lime

2 ounces (1 cup) unsweetened coconut flakes

Pineapple wedges, lime wedges, or maraschino cherries, for garnish (optional)

Simple syrup (optional)

day 1

Two to 8 hours before you plan to make your hard kombucha, prepare a yeast starter to ensure vigorous fermentation: Sanitize a 1-quart (or larger) jar and a spoon. In the jar, combine 2 cups of the pineapple juice, 2 ounces of the corn sugar, and the champagne yeast. Stir with the spoon until the sugar dissolves.

Cover with plastic wrap secured with a rubber band and let stand for 2 to 8 hours. You should see the plastic puff up and a layer of foam build on the surface of the liquid; both are signs of good fermentation.

When you're ready to make the hard kombucha, sanitize your fermenter, air lock, and whisk. In the fermenter, combine 10 ounces of the corn sugar, the remaining 2 cups pineapple juice, the everyday kombucha, and the yeast starter that you made. Whisk gently until the corn sugar is dissolved and the liquid is foamy on top, 30 to 60 seconds. Seal the fermenter, fill the air lock with sanitizer, and insert it into the fermenter.

Place the fermenter somewhere dark, slightly warm (70° to 80°F), and out of the way. You should start to see signs of fermentation (like bubbling in the air lock) within 24 to 48 hours.

Meanwhile, combine the oak cubes and rum in a small airtight container. Set aside for at least 5 days, shaking occasionally.

day 2	In a small, sanitized measuring cup, dissolve ½ teaspoon of the yeast nutrient in 2 tablespoons of warm water. Add to the fermenter, reseal, and replace the air lock. Swirl gently to distribute and return to your fermentation spot.
day 3	In a small, sanitized measuring cup, dissolve ¼ teaspoon of the yeast nutrient in 2 tablespoons of warm water. Add to the fermenter, reseal, and replace the air lock. Swirl gently to distribute and return to your fermentation spot.
day 4	In a small, sanitized measuring cup, dissolve the remaining ¼ teaspoon yeast nutrient in 2 tablespoons of warm water. Add to the fermenter, reseal, and replace the air lock. Swirl gently to distribute and return to your fermentation spot.
days 5 to 14	Active fermentation will peak around Day 5 and then mostly finish around Day 7. Anytime between Day 7 and Day 10, add the oak cube and rum mixture, citrus zest and juice, and coconut flakes to the fermenter and infuse for at least 3 days or up to 7 days. After infusing for 3 days, use a sanitized wine thief or measuring cup to scoop out a little liquid and give it a taste. Add more rum, orange juice, or lime juice until the desired flavor is reached. After adding flavorings, you may see renewed signs of fermentation, though less vigorous than originally. Once you see no more signs of fermentation, you can assume fermentation is complete (with kombucha, a few static bubbles floating in the air lock are fine as long as it's not actively bubbling). Wait another 24 hours to be safe, then proceed with bottling.
bottling day	When you're ready to bottle, sanitize a liquid measuring cup, spoon, large pot (1 gallon or larger), siphon, bottle filler, bottles, and caps. In the measuring cup, combine the remaining 1 ounce corn sugar with ½ cup of water and stir to dissolve. Pour this sugar water into the pot. Siphon the kombucha into the pot with the sugar water, leaving behind any solids. Attach the bottle filler to your siphon, transfer the hard kombucha into the bottles, and cap.
	Store somewhere cool, dark, and out of the way for 1 to 2 weeks to carbonate, and then refrigerate all the bottles. Since hard kombucha contains wild yeast, fermentation can sometimes continue after bottling and cause bottles to eventually burst if not refrigerated. Open bottles carefully outdoors or over a sink in case of gushing. Chill before enjoying and serve with garnish if desired. For a sweeter drink, add a splash of simple syrup before serving. Drink within 3 months.

4

HARD
Ciders

I'm a fan of all fermented beverages, of course, but hard cider holds a special place in my heart. With its snappy sweet-tart flavors and its bouncy, softly fizzing bubbles, cider is just such a *happy* drink. I love everything about it, from the way good apple juice smells as I pour it into the fermenter, to how easy it is to make (apple juice will practically ferment on its own if you just ask it politely), to the eager anticipation I feel as I pour it into a glass to take the first sip. Cider just makes me happy. I hope it makes you happy, too.

You don't need to press your own apples or buy gallons of fresh cider from the farmers' market to make your own hard cider. Good-quality store-bought juice will do the job just fine. My preference is for unfiltered apple juice (often labeled "cider") since it tends to have a bit more personality than the clear, ultrafiltered variety, but honestly, anything will work.

Unlike the other drinks in this book, there's plenty of sugar in apple juice already, so you don't need to add any more to make it boozy. That said, you're perfectly welcome to add more if you'd like to make a higher ABV cider, like the Fancy-Pants Imperial Cider (page 166), or if you want a specific flavor, like with the Apricot-Honey Hard Cider (page 150). Similarly, you don't need to add yeast nutrient for ciders to ferment well, though it doesn't hurt if you'd like extra insurance.

A hard cider made simply with apple juice is a treat all its own. But flavoring your hard cider with other fruits is when things really get fun. I have yet to meet a fruit that doesn't result in an excellent cider, from blueberries to mango. You can also swap some or even all the apple juice with another fruit juice. Many of these even have their own names, like perry (pear cider) and jerkum (plum, peach, or other cider made from stone fruit).

What to Expect: Cider fermentation tends to be more vigorous than the other kinds of drinks in this book. Especially if your fermenter is small, keep an eye on things during the first few super-active days of fermentation. Hard ciders also tend to finish pretty dry, so feel free to add a splash of simple syrup in your glass if you'd like.

basic HARD CIDER

AVERAGE ABV: 7 TO 8% ● **ABOUT 1 GALLON**

Ciders are straightforward and easy, and they're also great all on their own, even without adding other fruits or flavorings. Use an apple juice that you'd enjoy drinking and give it a taste once fermentation is complete. If you like it, drink it! If you want something more fruity, add fruit or other flavorings! You really can't go wrong.

 TRY THIS! Replace some or all the apple juice with any other juice of your choice, like pear, cherry, or pineapple. You can also juice your own mix of sweet and tart apples to make a unique blend.

1 gallon apple juice

½ teaspoon champagne yeast

Flavorings to taste

1 ounce (3 tablespoons) corn sugar

Simple syrup (optional)

---

day 1

Sanitize your fermenter, air lock, and whisk. In the fermenter, add the apple juice and champagne yeast. Whisk until the liquid is foamy on top, 30 to 60 seconds. Seal the fermenter, fill the air lock with sanitizer, and insert it into the fermenter.

Place the fermenter somewhere dark, slightly warm (70° to 80°F), and out of the way. You should start to see signs of fermentation (like bubbling in the air lock) within 24 to 48 hours.

---

days 2 to 14

Active fermentation will peak around Day 5 and then mostly finish around Day 7. Anytime between Day 7 and Day 10, add any flavorings to the fermenter and infuse for at least 3 days or up to 7 days.

After adding flavorings, you may see renewed signs of fermentation, though less vigorous than originally. Once you see no more signs of fermentation (like bubbles in the air lock), you can assume fermentation is complete. Wait another 24 hours to be safe, then proceed with bottling.

---

bottling day

When you're ready to bottle, sanitize a liquid measuring cup, spoon, large pot (1 gallon or larger), siphon, bottle filler, bottles, and caps. In the measuring cup, combine the corn sugar with ½ cup of water and stir to dissolve. Pour this sugar water into the pot.

Siphon the cider into the pot with the sugar water, leaving behind any solids. Attach the bottle filler to your siphon, transfer the hard cider into the bottles, and cap.

Store somewhere cool, dark, and out of the way for 1 to 2 weeks to carbonate, or for up to 3 months. Chill before enjoying. For a sweeter drink, add a splash of simple syrup before serving.

—

blueberry HARD CIDER

AVERAGE ABV: 7 TO 8% ● **ABOUT 1 GALLON**

If you've tried cider in the past and found it a little too tart, blueberry cider might change your mind. Blueberries are an easy-to-love berry to begin with, and here they add a noticeable sweetness to the cider. Blueberries also imbue the cider with a purple color that never fails to delight. If using fresh berries, crush them to break the skins before adding to the cider; frozen blueberries can be added as is.

 TRY THIS! Infuse your cider with orange zest, lemon zest, or chopped ginger to level up the flavors.

1 gallon apple juice

½ teaspoon champagne yeast

1 pound (about 3½ cups) fresh or frozen blueberries (lightly muddled if fresh)

1 ounce (3 tablespoons) corn sugar

Blueberries, for garnish (optional)

Simple syrup (optional)

---

day 1

Sanitize your fermenter, air lock, and whisk. In the fermenter, add the apple juice and champagne yeast. Whisk until the liquid is foamy on top, 30 to 60 seconds. Seal the fermenter, fill the air lock with sanitizer, and insert it into the fermenter.

Place the fermenter somewhere dark, slightly warm (70° to 80°F), and out of the way. You should start to see signs of fermentation (like bubbling in the air lock) within 24 to 48 hours.

---

days 2 to 14

Active fermentation will peak around Day 5 and then mostly finish around Day 7. Anytime between Day 7 and Day 10, add the blueberries to the fermenter and infuse for at least 3 days or up to 7 days.

After adding flavorings, you may see renewed signs of fermentation, though less vigorous than originally. Once you see no more signs of fermentation (like bubbles in the air lock), you can assume fermentation is complete. Wait another 24 hours to be safe, then proceed with bottling.

CONTINUED

bottling day

When you're ready to bottle, sanitize a liquid measuring cup, spoon, large pot (1 gallon or larger), siphon, bottle filler, bottles, and caps. In the measuring cup, combine the corn sugar with ½ cup of water and stir to dissolve. Pour this sugar water into the pot.

Siphon the cider into the pot with the sugar water, leaving behind any solids. Attach the bottle filler to your siphon, transfer the hard cider into the bottles, and cap.

Store somewhere cool, dark, and out of the way for 1 to 2 weeks to carbonate, or for up to 3 months. Chill before enjoying and serve with garnish if desired. For a sweeter drink, add a splash of simple syrup before serving.

two cherries HARD CIDER

AVERAGE ABV: 7 TO 8% ● **ABOUT 1 GALLON**

If one is good, two is better! This cider gets the benefit of both cherry varieties: juicy, purple-hued sweet cherries and mouth-puckering sour cherries. You can use fresh, pitted cherries if available, but frozen cherries are often easier to find and more affordable. You can even find bags of mixed sweet and sour cherries, so the work is already done for you!

 TRY THIS! Add a split vanilla bean or a few teaspoons of vanilla extract along with a whole cinnamon stick to make a "cherry pie" cider.

1 gallon apple juice

½ teaspoon champagne yeast

½ pound (about 1¾ cups) fresh or frozen pitted sweet cherries, roughly chopped

½ pound (about 1¾ cups) fresh or frozen pitted sour cherries, roughly chopped

1 ounce (3 tablespoons) corn sugar

Cherries, for garnish (optional)

Simple syrup (optional)

day 1

Sanitize your fermenter, air lock, and whisk. In the fermenter, add the apple juice and champagne yeast. Whisk until the liquid is foamy on top, 30 to 60 seconds. Seal the fermenter, fill the air lock with sanitizer, and insert it into the fermenter.

Place the fermenter somewhere dark, slightly warm (70° to 80°F), and out of the way. You should start to see signs of fermentation (like bubbling in the air lock) within 24 to 48 hours.

days 2 to 14

Active fermentation will peak around Day 5 and then mostly finish around Day 7. Anytime between Day 7 and Day 10, add the cherries to the fermenter and infuse for at least 3 days or up to 7 days.

After adding flavorings, you may see renewed signs of fermentation, though less vigorous than originally. Once you see no more signs of fermentation (like bubbles in the air lock), you can assume fermentation is complete. Wait another 24 hours to be safe, then proceed with bottling.

CONTINUED

bottling day

When you're ready to bottle, sanitize a liquid measuring cup, spoon, large pot (1 gallon or larger), siphon, bottle filler, bottles, and caps. In the measuring cup, combine the corn sugar with ½ cup of water and stir to dissolve. Pour this sugar water into the pot.

Siphon the cider into the pot with the sugar water, leaving behind any solids. Attach the bottle filler to your siphon, transfer the hard cider into the bottles, and cap.

Store somewhere cool, dark, and out of the way for 1 to 2 weeks to carbonate, or for up to 3 months. Chill before enjoying and serve with garnish if desired. For a sweeter drink, add a splash of simple syrup before serving.

mango HARD CIDER

AVERAGE ABV: 7 TO 8% ● **ABOUT 1 GALLON**

Mangoes and cider, you ask? Absolutely yes! I never would have thought to try mangoes in cider if I hadn't had a half bag of frozen mango chunks languishing in my overfull freezer and a gallon of hard cider in need of some love. I'm so glad for this discovery, because mango hard cider is delicious. It's juicy, fresh, and sweetly tart, just like a mango that's so ripe it's practically juice already.

 TRY THIS! Imbue this cider with a medley of tropical fruit flavors by using a mix of pineapple, kiwi, guava, and papaya along with the mango (keep the total amount of fruit to about 1 pound).

1 gallon apple juice	Strips of zest from 1 lemon
½ teaspoon champagne yeast	1 ounce (3 tablespoons) corn sugar
1 pound (about 3¼ cups) fresh or frozen mango chunks	Mango cubes, for garnish (optional)
	Simple syrup (optional)

day 1

Sanitize your fermenter, air lock, and whisk. In the fermenter, add the apple juice and champagne yeast. Whisk until the liquid is foamy on top, 30 to 60 seconds. Seal the fermenter, fill the air lock with sanitizer, and insert it into the fermenter.

Place the fermenter somewhere dark, slightly warm (70° to 80°F), and out of the way. You should start to see signs of fermentation (like bubbling in the air lock) within 24 to 48 hours.

days 2 to 14

Active fermentation will peak around Day 5 and then mostly finish around Day 7. Anytime between Day 7 and Day 10, add the mango and the lemon zest to the fermenter and infuse for at least 3 days or up to 7 days.

After adding flavorings, you may see renewed signs of fermentation, though less vigorous than originally. Once you see no more signs of fermentation (like bubbles in the air lock), you can assume fermentation is complete. Wait another 24 hours to be safe, then proceed with bottling.

bottling day

When you're ready to bottle, sanitize a liquid measuring cup, spoon, large pot (1 gallon or larger), siphon, bottle filler, bottles, and caps. In the measuring cup, combine the corn sugar with ½ cup of water and stir to dissolve. Pour this sugar water into the pot.

Siphon the cider into the pot with the sugar water, leaving behind any solids. Attach the bottle filler to your siphon, transfer the hard cider into the bottles, and cap.

Store somewhere cool, dark, and out of the way for 1 to 2 weeks to carbonate, or for up to 3 months. Chill before enjoying and serve with garnish if desired. For a sweeter drink, add a splash of simple syrup before serving.

pear-ginger HARD CIDER

AVERAGE ABV: 7 TO 8% ● **ABOUT 1 GALLON**

I love making ciders with pear juice, or "perries" as they're technically called, when I'd like more residual sweetness than we get with straight apple juice. Perries still end up fairly dry, but the pear flavor tends to read as sweeter than apple on our tongues. I usually buy pear juice in cartons, though you can make your own if you have a juicer.

 TRY THIS! If you tend to like your ciders a little sweet, try making any of the recipes in this chapter with half or all pear juice instead of apple.

½ gallon apple juice	1 ounce (3 tablespoons) corn sugar
½ gallon pear juice	Pear slices, for garnish (optional)
½ teaspoon champagne yeast	Simple syrup (optional)
½ ounce (2 tablespoons) roughly chopped ginger	

day 1

Sanitize your fermenter, air lock, and whisk. In the fermenter, add the apple juice, pear juice, and champagne yeast. Whisk until the liquid is foamy on top, 30 to 60 seconds. Seal the fermenter, fill the air lock with sanitizer, and insert it into the fermenter.

Place the fermenter somewhere dark, slightly warm (70° to 80°F), and out of the way. You should start to see signs of fermentation (like bubbling in the air lock) within 24 to 48 hours.

days 2 to 14

Active fermentation will peak around Day 5 and then mostly finish around Day 7. Anytime between Day 7 and Day 10, add the ginger to the fermenter and infuse for at least 3 days or up to 7 days.

After adding flavorings, you may see renewed signs of fermentation, though less vigorous than originally. Once you see no more signs of fermentation (like bubbles in the air lock), you can assume fermentation is complete. Wait another 24 hours to be safe, then proceed with bottling.

bottling day

When you're ready to bottle, sanitize a liquid measuring cup, spoon, large pot (1 gallon or larger), siphon, bottle filler, bottles, and caps. In the measuring cup, combine the corn sugar with ½ cup of water and stir to dissolve. Pour this sugar water into the pot.

Siphon the cider into the pot with the sugar water, leaving behind any solids. Attach the bottle filler to your siphon, transfer the hard cider into the bottles, and cap.

Store somewhere cool, dark, and out of the way for 1 to 2 weeks to carbonate, or for up to 3 months. Chill before enjoying and serve with garnish if desired. For a sweeter drink, add a splash of simple syrup before serving.

apricot-honey HARD CIDER

AVERAGE ABV: 7.5 TO 8.5% ● **ABOUT 1 GALLON**

Apricots are such a small but mighty fruit, and their intense sweet-tart flavor makes a cider that's hard to put down. This recipe uses dried apricots, but if you have been blessed with fresh, you can swap in 1 to 2 pounds. I also recommend seeking out California choice or Blenheim dried apricots for their almost candy-like flavor, but your cider will be delicious no matter what you get!

 TRY THIS! For a sweeter, even boozier drink, bump up the honey to a full cup. For a dessert wine, go for a full 2 cups and skip the corn sugar when bottling and serve this as a still (uncarbonated) cider.

1 gallon apple juice

6 ounces (½ cup) honey

½ teaspoon champagne yeast

6 ounces (1 cup) dried apricots (preferably California choice or Blenheim), roughly chopped

1 ounce (3 tablespoons) corn sugar

Fresh apricot slices, for garnish (optional)

Simple syrup (optional)

---

day 1

Sanitize your fermenter, air lock, and whisk. Heat 1 cup of the apple juice in the microwave until warm to the touch but not boiling, add the honey, and stir to dissolve. In the fermenter, add the apple-honey mixture followed by the remaining apple juice and then the champagne yeast. Whisk until the liquid is foamy on top, 30 to 60 seconds. Seal the fermenter, fill the air lock with sanitizer, and insert it into the fermenter.

Place the fermenter somewhere dark, slightly warm (70° to 80°F), and out of the way. You should start to see signs of fermentation (like bubbling in the air lock) within 24 to 48 hours.

---

days 2 to 14

Active fermentation will peak around Day 5 and then mostly finish around Day 7. Anytime between Day 7 and Day 10, add the apricots to the fermenter and infuse for at least 3 days or up to 7 days.

After adding flavorings, you may see renewed signs of fermentation, though less vigorous than originally. Once you see no more signs of fermentation (like bubbles in the air lock), you can assume fermentation is complete. Wait another 24 hours to be safe, then proceed with bottling.

bottling day

When you're ready to bottle, sanitize a liquid measuring cup, spoon, large pot (1 gallon or larger), siphon, bottle filler, bottles, and caps. In the measuring cup, combine the corn sugar with ½ cup of water and stir to dissolve. Pour this sugar water into the pot.

Siphon the cider into the pot with the sugar water, leaving behind any solids. Attach the bottle filler to your siphon, transfer the cider into the bottles, and cap.

Store somewhere cool, dark, and out of the way for 1 to 2 weeks to carbonate, or for up to 3 months. Chill before enjoying and serve with garnish if desired. For a sweeter drink, add a splash of simple syrup before serving.

mulled cranberry HARD CIDER

AVERAGE ABV: 7 TO 8% ● **ABOUT 1 GALLON**

This festive cider is a celebration of the season: it's a smooth balance of tart cranberry, soft citrus, and warm mulling spices, all in one sparkling glass. So, grab yourself a bag of cranberries as soon as you see them hit the stores in late October and make a batch (or two!) of this mulled cider for all your holiday party-going needs.

 TRY THIS! Skip the cranberries for a basic mulled cider, or skip the mulling spices to make a straightforward cranberry cider. You can also add 1 to 2 cups of honey at the start to make this more of a dessert wine. If you do, skip the corn sugar at bottling and serve this as a still (uncarbonated) cider.

1 gallon apple juice	5 whole allspice berries
½ teaspoon champagne yeast	2 whole star anise
½ pound (2 cups) fresh or frozen cranberries, roughly chopped	2 teaspoons vanilla extract
	1 ounce (3 tablespoons) corn sugar
Strips of zest and juice from 2 oranges	Cranberries, for garnish (optional)
2 cinnamon sticks	Simple syrup (optional)
5 whole cloves	

day 1

Sanitize your fermenter, air lock, and whisk. In the fermenter, add the apple juice and champagne yeast. Whisk until the liquid is foamy on top, 30 to 60 seconds. Seal the fermenter, fill the air lock with sanitizer, and insert it into the fermenter.

Place the fermenter somewhere dark, slightly warm (70° to 80°F), and out of the way. You should start to see signs of fermentation (like bubbling in the air lock) within 24 to 48 hours.

days 2 to 14

Active fermentation will peak around Day 5 and then mostly finish around Day 7. Anytime between Day 7 and Day 10, add the cranberries, orange zest and juice, cinnamon, cloves, allspice, star anise, and vanilla to the fermenter and infuse for at least 3 days or up to 7 days.

After 3 days of infusing, use a sanitized wine thief or measuring cup to scoop out a little liquid and give it a taste. Add more cranberries, orange juice, or any of the spices if a stronger flavor is desired.

CONTINUED

After adding flavorings, you may see renewed signs of fermentation, though less vigorous than originally. Once you see no more signs of fermentation (like bubbles in the air lock), you can assume fermentation is complete. Wait another 24 hours to be safe, then proceed with bottling.

bottling day

When you're ready to bottle, sanitize a liquid measuring cup, spoon, large pot (1 gallon or larger), siphon, bottle filler, bottles, and caps. In the measuring cup, combine the corn sugar with ½ cup of water and stir to dissolve. Pour this sugar water into the pot.

Siphon the cider into the pot with the sugar water, leaving behind any solids. Attach the bottle filler to your siphon, transfer the hard cider into the bottles, and cap.

Store somewhere cool, dark, and out of the way for 1 to 2 weeks to carbonate, or for up to 3 months. Chill before enjoying and serve with garnish if desired. For a sweeter drink, add a splash of simple syrup before serving.

sparkling rosé HARD CIDER

AVERAGE ABV: 7 TO 8% ● **ABOUT 1 GALLON**

Taste a glass of bona fide rosé and a glass of this cider side by side, and I honestly think you'd be hard-pressed to tell which one was the real deal. This fizzy pink cider is satisfyingly crisp and dry, but with a sweet fruitiness from the strawberries that lingers after each sip. Make it sparkling or still (just leave out the corn sugar at bottling), and serve it well chilled.

 TRY THIS! For a sweeter version, add a larger proportion of strawberries or swap the blackberries for blueberries. Or add more raspberries or blackberries if you like your rosé with more zip!

1 gallon apple juice

½ teaspoon champagne yeast

1 pound (16 to 20 medium) fresh or frozen strawberries (roughly chopped if fresh)

¼ pound (about 1 cup) fresh or frozen raspberries (lightly muddled if fresh)

¼ pound (about 1 cup) fresh or frozen blackberries (lightly muddled if fresh)

1 ounce (3 tablespoons) corn sugar

Simple syrup (optional)

---

day 1

Sanitize your fermenter, air lock, and whisk. In the fermenter, add the apple juice and champagne yeast. Whisk until the liquid is foamy on top, 30 to 60 seconds. Seal the fermenter, fill the air lock with sanitizer, and insert it into the fermenter.

Place the fermenter somewhere dark, slightly warm (70° to 80°F), and out of the way. You should start to see signs of fermentation (like bubbling in the air lock) within 24 to 48 hours.

---

days 2 to 14

Active fermentation will peak around Day 5 and then mostly finish around Day 7. Anytime between Day 7 and Day 10, add the strawberries, raspberries, and blackberries to the fermenter and infuse for at least 3 days or up to 7 days.

After adding flavorings, you may see renewed signs of fermentation, though less vigorous than originally. Once you see no more signs of fermentation (like bubbles in the air lock), you can assume fermentation is complete. Wait another 24 hours to be safe, then proceed with bottling.

CONTINUED

bottling day

When you're ready to bottle, sanitize a liquid measuring cup, spoon, large pot (1 gallon or larger), siphon, bottle filler, bottles, and caps. In the measuring cup, combine the corn sugar with ½ cup of water and stir to dissolve. Pour this sugar water into the pot.

Siphon the cider into the pot with the sugar water, leaving behind any solids. Attach the bottle filler to your siphon, transfer the hard cider into the bottles, and cap.

Store somewhere cool, dark, and out of the way for 1 to 2 weeks to carbonate, or for up to 3 months. Chill before enjoying. Serve with a few dashes of angostura bitters and garnish if desired. For a sweeter drink, add a splash of simple syrup before serving.

farmhouse saison HARD CIDER

AVERAGE ABV: 7 TO 8% ● **ABOUT 1 GALLON**

Golden saison, with its hints of lemon and earthy spice, is one of my favorite styles of beer, so of course I had to turn it into a cider! For this recipe, go for a Belgian ale yeast to bring some fruity and spicy character to the party, and then add lemon zest and juice, peppercorns, and coriander seeds to really amplify the flavors you'd find in a traditional saison.

 TRY THIS! If you have a favorite beer style, try it as a cider! Start by using the yeast that most closely matches the beer style, and then layer on hops, spices, or other flavoring ingredients to mimic what you taste in your pint glass.

½ gallon apple juice

½ gallon pear juice

½ teaspoon Belgian ale yeast, such as SafAle T-58

Strips of zest and juice from 1 lemon

1 teaspoon whole peppercorns

1 teaspoon whole coriander seeds

1 ounce (3 tablespoons) corn sugar

Lemon wedges, for garnish (optional)

Simple syrup (optional)

---

day 1

Sanitize your fermenter, air lock, and whisk. In the fermenter, add the apple juice, pear juice, and Belgian ale yeast. Whisk until the liquid is foamy on top, 30 to 60 seconds. Seal the fermenter, fill the air lock with sanitizer, and insert it into the fermenter.

Place the fermenter somewhere dark, slightly warm (70° to 80°F), and out of the way. You should start to see signs of fermentation (like bubbling in the air lock) within 24 to 48 hours.

---

days 2 to 14

Active fermentation will peak around Day 5 and then mostly finish around Day 7. Anytime between Day 7 and Day 10, add the lemon zest and juice, peppercorns, and coriander to the fermenter and infuse for at least 3 days or up to 7 days.

After adding flavorings, you may see renewed signs of fermentation, though less vigorous than originally. Once you see no more signs of fermentation (like bubbles in the air lock), you can assume fermentation is complete. Wait another 24 hours to be safe, then proceed with bottling.

---

bottling day

When you're ready to bottle, sanitize a liquid measuring cup, spoon, large pot (1 gallon or larger), siphon, bottle filler, bottles, and caps. In the measuring cup, combine the corn sugar with ½ cup of water and stir to dissolve. Pour this sugar water into the pot.

Siphon the cider into the pot with the sugar water, leaving behind any solids. Attach the bottle filler to your siphon, transfer the hard cider into the bottles, and cap.

Store somewhere cool, dark, and out of the way for 1 to 2 weeks to carbonate, or for up to 3 months. Chill before enjoying and serve with garnish if desired. For a sweeter drink, add a splash of simple syrup before serving.

cider OLD-FASHIONED

AVERAGE ABV: 8 TO 9% ● **ABOUT 1 GALLON**

When it comes to my go-to bar order, I'm an old-fashioned kind of lady through and through. This hard cider winks at that classic cocktail with an infusion of bourbon-soaked oak cubes and a twist of orange zest. Add a few dashes of Angostura bitters before sipping for perfection in a glass.

TRY THIS! If Manhattans are more your style, use rye whiskey instead of bourbon and add 6 ounces of cherries instead of the orange zest. Serve with a splash of sweet vermouth in the glass.

1 gallon apple juice	1 ounce (3 tablespoons) corn sugar
½ teaspoon champagne yeast	Angostura bitters, for serving
½ ounce oak cubes	Maraschino cherries and orange zest twists, for garnish (optional)
4 ounces (½ cup) bourbon	
Strips of zest from 1 orange	Simple syrup (optional)

day 1

Sanitize your fermenter, air lock, and whisk. In the fermenter, add the apple juice and champagne yeast. Whisk until the liquid is foamy on top, 30 to 60 seconds. Seal the fermenter, fill the air lock with sanitizer, and insert it into the fermenter.

Place the fermenter somewhere dark, slightly warm (70° to 80°F), and out of the way. You should start to see signs of fermentation (like bubbling in the air lock) within 24 to 48 hours.

Meanwhile, combine the oak cubes and bourbon in a small airtight container. Set aside for at least 5 days, shaking occasionally.

days 2 to 14

Active fermentation will peak around Day 5 and then mostly finish around Day 7. Anytime between Day 7 and Day 10, add the oak cube and bourbon mixture as well as the orange zest to the fermenter and infuse for at least 3 days or up to 7 days.

After 3 days of infusing, use a sanitized wine thief or measuring cup to scoop out a little liquid and give it a taste. Add more bourbon or orange zest until the desired flavor is reached.

After adding flavorings, you may see renewed signs of fermentation, though less vigorous than originally. Once you see no more signs of fermentation (like bubbles in the air lock), you can assume fermentation is complete. Wait another 24 hours to be safe, then proceed with bottling.

CONTINUED

bottling day

When you're ready to bottle, sanitize a liquid measuring cup, spoon, large pot (1 gallon or larger), siphon, bottle filler, bottles, and caps. In the measuring cup, combine the corn sugar with ½ cup of water and stir to dissolve. Pour this sugar water into the pot.

Siphon the cider into the pot with the sugar water, leaving behind any solids. Attach the bottle filler to your siphon, transfer the hard cider into the bottles, and cap.

Store somewhere cool, dark, and out of the way for 1 to 2 weeks to carbonate, or for up to 3 months. Chill before enjoying. Serve with a few dashes of angostura bitters and garnish if desired. For a sweeter drink, add a splash of simple syrup before serving.

hurricane HARD CIDER

AVERAGE ABV: 8 TO 9% ● **ABOUT 1 GALLON**

New Orleans's Hurricane cocktail sometimes gets a bad rap as an overly boozy, overly sweet drink sipped from plastic cups by tourists wandering the French Quarter. But if you skip the premade mixes, the Hurricane is a cocktail worthy of respect. This hard cider is a nod to the classic, made with aged rum, grenadine (pomegranate syrup), and citrus, with a splash of passion fruit flavoring for a tropical kick.

 TRY THIS! For an even fruitier flavor, replace 1 quart of the apple juice with 1 quart of pineapple juice.

1 gallon apple juice

½ teaspoon champagne yeast

½ ounce oak cubes

2 ounces (¼ cup) aged rum

2 ounces (¼ cup) grenadine

Strips of zest and juice from 1 orange

Strips of zest and juice from 2 limes

½ ounce (1 tablespoon) passion fruit flavoring, such as Brewer's Best

1 ounce (3 tablespoons) corn sugar

Maraschino cherries, for garnish (optional)

Orange or lime wedges, for garnish (optional)

Simple syrup (optional)

---

day 1

Sanitize your fermenter, air lock, and whisk. In the fermenter, add the apple juice and champagne yeast. Whisk until the liquid is foamy on top, 30 to 60 seconds. Seal the fermenter, fill the air lock with sanitizer, and insert it into the fermenter.

Place the fermenter somewhere dark, slightly warm (70° to 80°F), and out of the way. You should start to see signs of fermentation (like bubbling in the air lock) within 24 to 48 hours.

Meanwhile, combine the oak cubes and rum in a small airtight container. Set aside for at least 5 days, shaking occasionally.

---

days 2 to 14

Active fermentation will peak around Day 5 and then mostly finish around Day 7. Anytime between Day 7 and Day 10, add the oak cube and rum mixture, grenadine, citrus zest and juice, and passion fruit flavoring to the fermenter and infuse for at least 3 days or up to 7 days.

After 3 days of infusing, use a sanitized wine thief or measuring cup to scoop out a little liquid and give it a taste. Add more rum, grenadine, orange juice, lime juice, or passion fruit flavoring until the desired flavor is reached.

CONTINUED

After adding flavorings, you may see renewed signs of fermentation, though less vigorous than originally. Once you see no more signs of fermentation (like bubbles in the air lock), you can assume fermentation is complete. Wait another 24 hours to be safe, then proceed with bottling.

bottling day

When you're ready to bottle, sanitize a liquid measuring cup, spoon, large pot (1 gallon or larger), siphon, bottle filler, bottles, and caps. In the measuring cup, combine the corn sugar with ½ cup of water and stir to dissolve. Pour this sugar water into the pot.

Siphon the cider into the pot with the sugar water, leaving behind any solids. Attach the bottle filler to your siphon, transfer the hard cider into the bottles, and cap.

Store somewhere cool, dark, and out of the way for 1 to 2 weeks to carbonate, or for up to 3 months. Chill before enjoying and serve with garnish if desired. For a sweeter drink, add a splash of simple syrup before serving.

fancy-pants IMPERIAL CIDER

AVERAGE ABV: 8 TO 9% ● **ABOUT 1 GALLON**

Sure, straight-up hard cider is good, but wouldn't the party be more fun if we made it with maple syrup? And while we're at it, why don't we add some Calvados (apple brandy) to really seal the deal? The answer to both questions is obviously yes, and the result is a rich, semisweet imperial cider that certainly doesn't mind being a little bit extra.

 TRY THIS! Swap the maple syrup for honey or brown sugar, and swap the Calvados for rum, bourbon, or cognac.

1 gallon apple juice

5½ ounces (½ cup) maple syrup

½ teaspoon champagne yeast

½ ounce oak cubes

2 ounces (¼ cup) Calvados or apple brandy

1 ounce (3 tablespoons) corn sugar

Simple syrup (optional)

---

day 1

Sanitize your fermenter, air lock, and whisk. Heat 1 cup of the apple juice in the microwave until warm to the touch but not boiling, add the maple syrup, and stir to dissolve. In the fermenter, add the apple-maple mixture followed by the remaining apple juice and then the champagne yeast. Whisk until the liquid is foamy on top, 30 to 60 seconds. Seal the fermenter, fill the air lock with sanitizer, and insert it into the fermenter.

Place the fermenter somewhere dark, slightly warm (70° to 80°F), and out of the way. You should start to see signs of fermentation (like bubbling in the air lock) within 24 to 48 hours.

Meanwhile, combine the oak cubes and Calvados in a small airtight container. Set aside for at least 5 days, shaking occasionally.

---

days 2 to 14

Active fermentation will peak around Day 5 and then mostly finish around Day 7. Anytime between Day 7 and Day 10, add the oak cube and Calvados mixture to the fermenter and infuse for at least 3 days or up to 7 days.

After adding flavorings, you may see renewed signs of fermentation, though less vigorous than originally. Once you see no more signs of fermentation (like bubbles in the air lock), you can assume fermentation is complete. Wait another 24 hours to be safe, then proceed with bottling.

---

bottling day

When you're ready to bottle, sanitize a liquid measuring cup, spoon, large pot (1 gallon or larger), siphon, bottle filler, bottles, and caps. In the measuring cup, combine the corn sugar with ½ cup of water and stir to dissolve. Pour this sugar water into the pot.

Siphon the cider into the pot with the sugar water, leaving behind any solids. Using a sanitized wine thief or measuring cup, scoop out a little liquid and give it a taste. Add more Calvados if desired. Attach the bottle filler to your siphon, transfer the hard cider into the bottles, and cap.

Store somewhere cool, dark, and out of the way for 1 to 2 weeks to carbonate, or for up to 3 months. Chill before enjoying. For a sweeter drink, add a splash of simple syrup before serving.

Resources

SUPPLIES

Brooklyn Brew Shop (brooklynbrewshop.com): This is a good source for kombucha scobys. They have a full range of fun 1-gallon brewing kits that you can try.

Kombucha Kamp (kombuchakamp.com): This site is another fantastic source for scobys, kombucha-making kits, and other supplies, plus you'll find a wealth of knowledge if you're interested in diving more deeply into the world of kombucha.

MoreBeer! (morebeer.com): This is an excellent source for the full range of homebrewing supplies, from fermenters to bottles.

Northern Brewer (northernbrewer.com): Another excellent source for homebrewing supplies, and the best place to find the Little Big Mouth Bubbler, my favorite fermenter.

PerfectWerks (perfectwerks.com): If you get tired of bottling, pick up one of this company's mini-kegs. I've had mine for years and love how easy it is to use.

The Spice House (thespicehouse.com) or Mountain Rose Herbs (mountainroseherbs.com): These are my preferred vendors for whole spices like peppercorns and coriander seed.

TeaSource (teasource.com): I buy all my specialty loose-leaf teas from this online store, like the hibiscus for making Agua de Jamaica Hard Iced Tea (page 87).

BOOKS

The Art of Fermentation by Sandor Katz: This book explores the world of fermented foods and drinks from every possible angle and is essential for any fermentation enthusiast's bookshelf.

The Big Book of Kombucha by Hannah Crum and Alex LaGory: If you're interested in kombucha, this is *the* book to get. It covers every aspect of kombucha brewing in great detail and offers a lot of advice and trouble-shooting suggestions for new and veteran brewers alike.

The Drunken Botanist by Amy Stewart: This book covers all the herbs, spices, and botanicals that have ever been used to make alcoholic beverages in delightfully nerdy detail. I find this book to be a source of inspiration when crafting new recipes.

The Farmsteady Guide to Kombucha by Erica Shea and Stephen Valand: From the founders of Brooklyn Brew Shop and Farmsteady, this is another great intro book for new kombucha makers. It also has dozens of fun, fizzy, flavorful recipes to try.

The New Cider Maker's Handbook by Claude Jolicoeur: If you fall in love with making hard cider, this book will be an incredible resource for you. It's geared toward professional cider makers, so it gets a little technical at times, but it offers a wealth of knowledge.

COMMUNITY

American Homebrewers Association (homebrewersassociation.org): Though primarily devoted to home beer brewing, this organization is a great resource for all home fermentation projects.

Homebrew Talk (homebrewtalk.com): This is an open forum for homebrewers and fermenters of all stripes. It's a great place to connect with fellow enthusiasts and get help troubleshooting issues with your projects.

Acknowledgments

No project like this can be accomplished without the help of many. I am so grateful to each and every individual who touched this book.

To my husband, Scott: Thank you for your unwavering support and also for letting me sacrifice half of our kitchen to this project for over a year.

To my Minnesota family and my Massachusetts family: Thank you for all the articles and newspaper clippings and little moments of encouragement you sent me over the course of this project!

To my brother, Andy: You get a particular shout-out for being my most steadfast brewing buddy. I've loved our virtual brew days together and our endless chat threads delving into the mysteries of fermentation. I only wish we lived closer so we could share more of our homebrews together.

To my agent, Angela Miller: I so appreciate having you in my corner. Thank you for supporting my work all these years.

To my team at Ten Speed—my editor, Claire Yee; my designer, Betsy Stromberg; and my copyeditor, Heather Rodino: We did it again! This book was absolutely a team effort, and I'm so grateful to have had your experience, wisdom, and enthusiastic support throughout this process.

To my photography team, Erin Kunkel, Vanessa Solis , and Kyle Emery-Peck; food styling team, Abby Stolfo and Chelsea Lopker; and prop stylist, Nissa Quanstrom: You are legit wizards. Thank you for making all these drinks look drop-dead gorgeous, for being so fun and supportive on set, and for convincing me that a few cubes of ice in my drink is, in fact, quite fabulous.

To my superstar team of recipe testers—Ana Kolpin, Andy Christensen, Anthony Betts, Charles Thresher, Christine Goerss-Barton, Chuck Grimmett, Danielle Hanlon, David Renaud, Diane Tani, Gilbert Seward, Jaclyn Mandart, Laney Vela, Mark Beahm, Mary Coughlin, Megan Scott, Patti Bradley, Tiffany Buckwalter, Traci Downing, and Zac Dillon: Thank you for helping kick the tires on all these recipes and kindly pointing out when and where the instructions made no sense. This book is a thousand times better thanks to your feedback.

To my lovely lady squad—Heather Pagano, Heather Brady, and Sara Rosenberg: You all deserve a medal for listening to the countless voice messages I left in our group chat over the course of this book and for always responding with steadfast encouragement. You're the best friends a gal could ask for.

To my St. Leo neighborhood gang: Thank you for giving me your honest opinions on all the brews I brought to our Friday pizza nights and for going down all the nerdy recipe/ingredient/fermentation rabbit holes with me. Also thank you for your enthusiastic cheerleading throughout this whole project.

To my Simply Recipes, The Spruce Eats, and Dotdash Meredith crew: Thank you for indulging my many moments of spaciness while working on this book and my occasional digressions into fermentation topics during meetings. It meant so much to know you were cheering me on from the sidelines.

Finally, I want to acknowledge the work of the Pink Boots Society, a nonprofit organization supporting women and nonbinary individuals in the fermented/alcoholic beverage industry. A portion of my advance for this book has been donated to this organization. To learn more, visit www.pinkbootssociety.org.

Index

Ten Speed Press
An imprint of the Crown Publishing Group
A division of Penguin Random House LLC
tenspeed.com

Typefaces: Latinotype's Campeche, Melvastype's Santeli, and OGJ Type Design's Shapiro Base

Library of Congress Cataloging-in-Publication Data
Names: Christensen, Emma, author. | Kunkel, Erin, photographer. Title: Hard seltzer, iced tea, kombucha, and cider : how to make your own boozy fermented drinks at home / Emma Christensen ; photography by Erin Kunkel. Identifiers: LCCN 2024016291 (print) | LCCN 2024016292 (ebook) | ISBN 9780593835777 (hardcover) | ISBN 9780593835784 (ebook) Subjects: LCSH: Fermented beverages. | Fermentation. | LCGFT: Cookbooks. Classification: LCC TP371.44 .C59 2025 (print) | LCC TP371.44 (ebook) | DDC 641.2—dc23/eng/20240823
LC record available at https://lccn.loc.gov/2024016291
LC ebook record available at https://lccn.loc.gov/2024016292

Hardcover ISBN: 978-0-593-83577-7
eBook ISBN: 978-0-593-83578-4

Printed in China

Acquiring editor: Julie Bennett | Project editor: Claire Yee | Production editor: Serena Wang
Art director and designer: Betsy Stromberg | Production designers: Mari Gill and Faith Hague
Production manager and Prepress color manager: Jessica Heim
Food stylist: Abby Stolfo | Food stylist assistant: Chelsea Lopker
Prop stylist: Nissa Quanstrom | Prop stylist assistant: Katherine Mohun
Photo assistants: Vanessa Solis and Kyle Emery-Peck
Copyeditors: Rachel Holzman and Heather Rodino
Proofreaders: Erica Rose and Kate Bolen | Indexer: Jay Kreider
Publicist: Lauren Chung | Marketer: Joey Lozada

10 9 8 7 6 5 4 3 2 1

First Edition

"No one is better at demystifying and simplifying kitchen-scaled fermentation than Emma Christensen. Her latest fermented tour de force is equal parts educational and entertaining, providing a recipe-driven road map to the rapidly expanding drinking world."

—JOSHUA M. BERNSTEIN, author of *The Complete Beer Course*

"Emma Christensen has done it again! She has taken what seems to be a challenging process and made it simple and fun. With clear instructions and beautiful photos, Emma makes the trendiest alcoholic beverages a snap to whip up in the comfort of your own home. Plus, the flavor combinations are fresh, fruity, and fantastic! If you've ever wanted to try your hand at fermentation, yet were afraid to start, this is an easy way to learn this ancient art."

—HANNAH CRUM, coauthor of *The Big Book of Kombucha,* founder of Kombucha Kamp, and cofounder of Kombucha Brewers International